BREEDING CAGE AND AVIARY BIRDS

**Other Fine Howell Books for
the Bird Owner's Library**

All About the Parrots, *Arthur Freud*
Bird Owner's Home Health and Care Handbook,
Gary A. Gallerstein, D.V.M.
The Complete Cockatiel, *Matthew M. Vriends*
Popular Parrots, *Matthew M. Vriends*

The modern Canary was developed from a wild bird. Skillfully applied genetics, carried on for several centuries, has resulted in the great variety of distinct types at the present time. *Photo by Author*

Breeding Cage
and
Aviary Birds

by

Dr. Matthew M. Vriends

First Edition — First Printing

1985

HOWELL BOOK HOUSE INC.
230 Park Avenue
New York, N.Y. 10169

Library of Congress Cataloging in Publication Data

Vriends, Matthew M., 1937-
  Breeding cage and aviary birds.

  Bibliography: p. 191
  1. Cage-birds—Breeding.   2. Birds, Ornamental—
Breeding. I. Title.
SF461.8.V75   1984   636.6′86   84-22415
ISBN 0-87605-821-7

Contents

BREEDING CAGE AND AVIARY BIRDS

Dr. Matthew M. Vriends

About The Author

FOR MATTHEW M. VRIENDS, an abiding interest in the animal kingdom is a family affair that spans the generations.

A native of Eindhoven in the southern Netherlands, Dr. Vriends was strongly influenced by his father's example and involvement in the sciences. The elder Vriends was a celebrated writer and respected biology teacher. An uncle also loomed large in the Dutch scientific community, and the Natural History Museum at Asten, which annually welcomes over 200,000 visitors, is named in his honor.

Dr. Vriends vividly remembers the field trips he shared with both father and uncle and the "mini menugerie" maintained in the Vriends family home. The facilities for keeping and observing flora and fauna included a pair of large aviaries housing over 50 tropical bird species. A source of particular pride is the fact that many *first breeding results* came about in the Vriends family aviary.

Dr. Vriends' first published material appeared in magazines while he was still in high school. He wrote and illustrated his first bird book at age 17—an amazing achievement that got young Vriends officially named the youngest biologist with a published work to his credit. This book was also an unqualified publishing success with more than six reprints and sales in excess of 40,000 copies!

During his University career, Matthew Vriends continued to publish literary essays, poetry, short stories, a number of fine bird books and even a novel that helped finance his education.

After graduation he worked as a high school teacher, but eventually left education to devote more time to the serious study of ornithology. His work took him to some of the world's most exotic ports of call in Africa, South America, Indonesia and Australia. And it was in Australia—home of many of the world's most beautiful and unusual bird species—that Matthew Vriends became fascinated with native parrots, parakeets and grassfinches. He remained in Australia from 1964 through 1967 absorbed in study and continuous publication of ornithological subjects.

A number of the books published during these years came to the attention of Dr. Franz Sauer, world-famous ornithologist, who succeeded in persuading Dr. Vriends to come to the University of Florida at Gainesville. Here Vriends worked with world-renown biologists and to broaden his horizons, worked at the veterinary science and medical laboratories. One credit of which the author is particularly proud is having been allowed to work on the influenza virus research being conducted at that time.

Matthew Vriends earned his American doctorate in 1974 with a thesis on the Australian Masked Grassfinch (*Poephila personata*), and returned to Holland following the completion of his studies. Some time later he crossed the ocean again to take a position as the senior ornithology editor with a large American publishing firm until family concerns necessitated yet another return to the Netherlands in 1980. Vriends remained based in Holland until mid-1983 where he worked as an educator with additional interests in publishing and writing. He and his family now make their home on Long Island.

He remains an avid world-traveler, and, with his wife Lucy and daughter Tanya, regularly visits various countries of the world to observe the local wildlife close-up. His extensive travel also included an annual journey to the United States while he resided in Holland. He continues as well-known and respected by the American avicultural community as by the Dutch.

As his father did, Matthew Vriends maintained a large, varied collection of animals in his home for both enjoyment and study before coming to the U.S. Fish, hamsters, gerbils, mice, rats, guinea pigs, turtles, dogs and, of course, birds—some 80 different species—constituted the 1981 Vriends family menagerie. Happily, it appears that Tanya Vriends will be the third generation biologist/aviculturist in the family as she joins her parents with great enthusiasm in their interest.

Dr. Vriends generously shares his expertise in various ways. A popular international judge, he frequently officiates at bird shows in many countries. In Holland he was the host of a weekly radio program and conducted seminars on birds and other animals in conjunction with trade show appearances during his American visits. His greatest fame has come through his writing and the helpful information he has imparted to pet owners and fanciers far and wide.

Currently Dr. Vriends is the author of over 80 books, in three languages, on birds, mammals, bees, turtles and fish, and over 1000 articles that have appeared in American and European magazines.

This remarkably prolific individual also enjoys music, painting, sketching, photography, tennis and gardening during rare moments of leisure when his attention is not directed to the natural sciences.

Dr. Matthew M. Vriends' accomplishments are like those of few others. By his varied activities in his chosen field, he has enlarged the body of knowledge for scientists, naturalists, diverse fancier groups and pet lovers around the world. His international celebrity is earned through more than thirty years of education, achievement and enthusiastic devotion to science and aviculture.

Saffron Finch *(Sycalis flaveola)*; 6 inches (15 cm). Habitat: southern Brazil, Venezuela to New Granada. The female is paler beneath and duller above.　　　　　*Photo by Author*

Rothschild's Mynah *(Leucopsar rothschildi)*; approx. 10 inches (26 cm). Habitat: Bali (Indonesia).
Photo by Author

Foreword

BREEDING BIRDS in a cage or aviary is currently a very popular hobby. People like the challenge of working with birds because it requires special attention and expertise that generally takes years to develop.

I have personally been breeding and raising birds of various species for more than a quarter century, so can draw upon considerable practical experience. While my own background is an important part of this book, I have also brought in the parallel experiences of many aviculturist friends throughout the world.

My intent in this book is to be more practical than scientific. I am aware that, biologist as I am, I have not resisted adding a paragraph or two that strictly speaking is not essential to the practical raising of birds. I hope these extra insights add to the reader's general background and enjoyment.

My approach is deliberately not "tropical." I use examples of many North American bird species to simplify the discussion. If the reader wishes, he can check me out through personal observations, which naturally couldn't be made with tropical and subtropical birds.

I have tried to let each topic in the chapter "Natural Bird Breeding" stand on its own as much as possible. This results in a certain amount of repetition, but—given the intent to make this book a handy reference—I believe this is justified.

I have given considerable space to the hookbills, many of which are now endangered species. I believe that as aviculturists, we can and will play a very important role in the preservation of these and other endangered avian species. I don't mean to show any personal preference to parrots and parakeets, even though I dwell on them at length. This emphasis is generated more or less out of necessity. Extensive attention is also devoted to Budgerigars, since they are far and away the most popular pet birds in America.

I don't make a claim to absolute wisdom. When I present my way of raising young, I don't mean to say this is the only way, or the only good way. The text which follows is completely subject to any valid corrections and additions, which may become necessary.

I hope this book can make a contribution, small as it may be, to the preservation of our natural environment, which is under threat from all sides. I believe that the serious bird owner and breeder is no threat to the survival of birds in Nature in any part of the world—no matter what many bird and animal protectionists say. The opposite is true. Efforts of serious aviculturists in the past few years have saved a number of bird species from total extinction. This is an observation well worth making.

Let us then be informed participants in a creative effort that will and should be dynamic. Good luck.

MATTHEW VRIENDS

Acknowledgments

I AM INDEBTED to the many aviculturists and ornithologists who replied so helpfully to my requests for information, especially my friends Mr. and Mrs. Remi Ceulecrs and Paul Leysen, Herentals, Belgium; Mr. Paul Kwast, editor of *Vogelvreugd* (Holland), and the management of Elsevier and Keesing, Publishers in Amsterdam, Holland. My grateful thanks are due to Mr. Max B. Heppner, Takoma Park, Maryland for his skilled editing and to Mr. Seymour N. Weiss, my editor at Howell Book House Inc., New York, for his ready help in all matters relating to this book.

Thanks to my wife, Lucy Vriends-Parent, for her invaluable assistance and patience during the preparation of the text; without her this book could never have been written. All the opinions and conclusions expressed in the following pages are my own, however, and any errors must be my own responsibility.

M.M.V.

Aurora Finch *(Pytilia phoenicoptera)*; approx. 4½ inches (11.5 cm). Habitat: west and central Africa. The female is paler and browner. *Photo by Author*

Cordon Blue, female (Uraeginthus bengalus); approx. 4½ inches (11.5 cm). Habitat: east Africa. The male has a red ear-patch. *Photo by Author*

1

Introduction—
A Comprehensive
Overview

THE EXPRESSION "bird lover" doesn't tell the whole story. I strongly suspect that many bird fanciers have a special love for the eggs, the nests, and the young of their birds. So a discussion of these interesting topics is most worthwhile. Let's raise and answer some general questions and use them as an introduction. Later we can examine some special cases.

All birds are egg layers. We also know they're not the only animls that lay eggs. Fish, reptiles, and other genera also reproduce in this fashion.

Eggs

There is something special about bird eggs, however. The calcareous shell, for example, is quite porous to make possible the exchange of gases.

Of more interest is the number of eggs birds lay in a clutch. In Nature, this number can vary greatly. The low end of the range is held by single-egg-laying birds, including several large sea birds, the Arctic Tern (*Sterna paradisaea*), and a type of swallow from Borneo. The upper end of the range—10 to 20 eggs—is exemplified by the Golden-crowned Kinglet (*Regulus satrapa*), which lays up to 10; the Mallard (*Anas platyrhynchos*), which lays up to 16; and the Ring-necked Pheasant (*Phasianus colchicus*), which lays up to 15.

Is that happenstance? No. The number of eggs a bird lays is of key importance for the preservation of a species. The single-egg-laying bird survives that way by selecting its brooding site in an inaccessible place. That Borneo Swallow, for example, constructs a cup-shaped nest out of mud at the tip of a thin, supple twig. Most large sea birds lay their eggs on an inaccessible rock, often hiding them somewhat in a chink. Their enemies stand practically no chance of finding either eggs or young.

Compare that, if you will, with pheasants or ducks, which have their breeding sites on the bare ground. So to speak, everything is all laid out for any egg-eating enemy. So we can arrive at a general rule: "The greater the danger, the larger the number of eggs."

Does that mean that females of a certain species always lay the same number of eggs? Generally, that is so. A Bobwhite (*Colinus virginianus*) will lay 7 to 20 eggs; a Herring Gull (*Larus argentatus*) - 3, perhaps 4; a Long-billed Curlew (*Numenius americanus*) - 4, sometimes 3 or 5; a Lapwing (*Vanellus vanellus*) - 4, sometimes 5; and a Ring-necked Duck (*Aythya collaris*) - 7 to 13 and sometimes more. (In the case of very large clutches, however, it may be due to two females using the same nest.) A Sharp-shinned Hawk (*Accipiter striatus*) generally lays 5 eggs; a Robin *(Turdus migratoris),* 3 to 5; and a House Sparrow (*Passer domesticus*), 4 to 6.

Sometimes there is a fairly large range in a species (like the Ring-necked Duck), and there always seems to be room for expansion. That's evident to anyone who has gathered Lapwing eggs, an activity many Europeans engage in. The gatherers (at least the good ones) make sure that the Lapwing generously exceeds her normal clutch (4 eggs). Not to speak of the domestic chicken, which has been adapted to lay hundreds of eggs per year.

In a bird magazine, I once read about the fantastic clutch of 32 eggs laid by a Zebra Finch. Another ornithological magazine, *The Auk,* had an article years ago about the persistent egg-laying marathon of a woodpecker, which managed to lay 71 eggs in 73 days. (Also refer to *This Fascinating Animal World* by A. Devoe.)

Yes, we have heard some fabulous tales on this score from pigeon fanciers and from aviarists. The purpose of this discussion is not to try for new records with our own aviary birds, however. Remember though, pressure to produce more eggs tires the females, with a definite chance that future breeding plans will be ruined or that the females may die from egg binding. The breeder can still try for a modest and controlled step-up in the size of the clutch—at least in some cases.

There is more to say about eggs—for example, about their color. Some attribute coloration to mimicry, meaning a protective camouflage. But that is not always the case. The most accepted theory is connected with the fact that the evolutionary ancestors of our birds were reptiles. Most likely, the first bird eggs were a whitish yellow. As nestbuilding developed, the coloration of eggs developed with it to produce a camouflage or mimicry color. This took place in the course of thousands of years. A hole breeder, for example the Downy Woodpecker (*Dendrocopos pubescens*), lays shiny white eggs, but this eye-catching color doesn't pose a threat because the eggs are not noticeable in their pear-shaped hole in a tree.

The situation is quite different with the eggs of the Long-Billed Curlew (*Numenius americanus*). This bird builds its nest between heather bushes, dune thistles, grass or other low-growing ground cover. The nest is insignificant and the olive or brownish green eggs are not evident, camouflaged as they are with brown, brownish-yellow, and purplish-grey flecks. That is why the clutch isn't too big—just 3 to 5 eggs. The color of the eggs evolved toward brownish green with brown spots for security reasons.

What about all those clear blue, pink, dark blue, green and other bright egg colors of birds nesting in trees? The coloration of these eggs developed while the birds were still breeding on solid ground, and the color served as a protection. When the birds took to trees for nesting, that protective color was no longer necessary. If you observe in Nature, you will notice that eggs of a particular bird species that presently nests in trees can differ in color and markings. It is fairly safe to conclude that these eggs are slowly but surely changing color. When they will all ultimately become white is, of course, impossible to say.

Nests

It seems tempting to answer with a simple, *yes,* when the question arises whether all birds build nests. But this is not true in every case. Many ground breeders construct no nest to speak of, or a very simple one. Beach birds, like plovers, twist out a small hole in the sand, in which they lay their eggs. Some birds have the artistic tendency to decorate the edge of the nest with a single seashell. This may have a sheer practical value to help identify the nest and help locate it. My own research has tended to confirm this explanation.

Even if beach birds are still considered to have nests, it is certain that brood parasites, like the well-known cuckoo, certainly don't build nests. It seems quite strange that cuckoos entrust the care of their young to others. But further thought leads us to the recognition that in Nature literally everything depends on everything else. The true relationship between a cuckoo hen and the foster mother she selects for her chicks isn't precisely known, but ornithological studies show promise, especially in comparing the lifestyle of the cuckoo with that of the cowbird.

In short, most birds build nests, pretty ones or less pretty ones. Ground breeders tend to build quite simply. The Lapwing twists a depression into the grass and a Quail tries to form the depression into a kind of tunnel. On the other extreme are the attractive and complex nests of birds like the Long-tailed Tit (*Aegithalos caudatus*) of Europe, the Baltimore Oriole (*Icterus galbula*) of North America, or the weaver birds of Africa. Weavers are known to fill entire acacia trees with pear-shaped nests, which, when viewed from a distance, resemble fruit. Nesting trees are then sometiomes covered with a common roof.

Nests are built at varying heights: on the ground, nests of Lapwings or the Horned Lark (*Eremophila alpestri*) are found; in low bushes or hedges, the Mocking Bird (*Mimus poluglottos*), the Hermit Thrush (*Catharus guttatus*) and the various finches built their nests; the Cooper's Hawk (*Accipiter cooperii*) and the Black-Billed Magpie *(Pica pica)* prefer to nest high in the trees; in hollow trees and towers, parrots, owls, and wood-peckers often set up housekeeping; hollows at the water's edge or in rabbit holes are the choice of the Kingfisher (*Megaceryle torquata*), the Sand Martin (*Riparia riparia*) and the European and Asiatic Shelduck (*Tadorna tadorna*).

Nest elevation can vary at times, as with the House Finch *(Carpo-dacus mexicanus)* and the Robin *(Turdus migratoris)*. The ground cover can have an influence, as with the Swamp Sparrow *(Melospiza georgiana),* which builds in the rushes; the Blue-winged Warbler *(Vermivora pinus),* which builds in swamp growth; and the Greater Swamp Duck *(Aythya marila nearctica)* and the European Aquatic Warbler *(Acrocephalus paludicola),* which build in the sedges. Nest elevation also can depend on factors like open topography, as with the Great Grey Shrike *(Lanius excubitor)*; or flowing water, as with several types of swallow.

When traditional sites and natural vegetation disappear through agricultural and other human encroachment, birds can disappear alto-gether, or nearly so. One can see a sharp decline in woodpeckers, Red-breasted Nuthatches (*Sitta canadensis*), tits, and Brown Creepers (*Certhia familiaris*) when there is a shortage of sick and hollow trees. Similarly, the Hoopoe (*Upupa epops*), the Night Heron (*Nycticorax nicticoral*), and the European Golden Plover (*Pluvialis apricaria*) are disappearing because of agricultural development.

On the other hand, birds *can* adapt to new circumstances, satisfied with novel nesting places. We find nests in mail boxes, old pumps, empty flower pots, and birdhouses. Birds will even change construction materials, like candy wrappers, hair, and wool—plucked from people's coats and shawls.

When do birds build? Most observations find them at work in the morning, although this also differs among species.

Another question is, which partner builds the nest? In most cases, it is the female, which also chooses the location of the nest. But often both the male and the female work at nest building. Generally, one can assume that males and females which are identical in coloration build nests together. In

Red-legged Honey Creeper, male (*Cyanerpes cyaneus*); 4-5 inches (10-13 cm). Habitat: Cuba and southern Mexico to Bolivia. Can also be found on Jamaica. The female is dark green, lighter on the head; the underside is also light green.   *Photo by Author*

Long-tailed Grassfinch *(Poephila acuticauda)*; approx. 4½ inches (11.5 cm). Habitat: the Northern Territory and northwestern Australia. The female usually has a smaller bib.
Photo by Author

cases where this isn't so, the female builds the nest, but the male often brings in construction material.

A final question—do birds reuse old nests? Here again there is no definitive answer. It is well-known that storks and swallows really do like to return to their old nests, and their offspring frequently do, too. Other birds, like the House Finch, the Rock Dove (*Columba livia*), the Crow, and the Robin, like to build new nests in the immediate vicinity of their old ones.

Most birds, however, use a nest only once, particularly birds like pheasants, partridges, quails, ducks, and similar species, whose young run out immediately and so don't use the nest extensively. There are species that construct separate hiding nests and sometimes even trial nests. That's true for the popular fancy bird, the Zebra Finch, as well as the common wren. Thus, a single brood may require three or four nests.

In a few cases, birds use a nest for successive broods, as with the sparrow, the pigeon, and jackdaw. The popular aviary dove, the Diamond Dove (*Geopelia cuneata*) is another good example. It makes "improvements." At first, the nest is a tumbledown creation, but then it develops into a complex whole, flicked together. It is not too attractive, but it does the job.

That leads us straight to the question of why and how birds brood. Or, do all birds brood?

Brooding Behavior

There comes a time when a certain series of instincts that preserves the species takes hold. That is the start of the brooding season for birds. Inside them, something changes. People often say, birds are broody or ready to brood. Old bird lovers used to say, "The birds are running a fever."

What's the true process? Yes, the elevation of temperature at brooding is a reality, but it involves only certain limited locations of the body, called brood spots. While sitting on her eggs, the female presses these spots against the eggs.

Not all birds have the brood spots at the same location. That's why all birds don't sit on their eggs the same way. Take a look at the comfortable, broadly stretched-out domestic hen. Before settling on her eggs, she first arranges nest and eggs to make them fit her form pleasurably. Then, compare the super-nervous Lapwing, which presses breast on eggs and sits with body steep and pointed—seemingly as if she has just a moment's time and has urgent business elsewhere. Compare also the deeply settled House Finch with the airily-scarily sitting Zebra Finch. Or, compare the Robin, which ducks down deep into her nest, with the Cormorant, sitting royally on her throne. These positions are wholly dictated by the location, size and intensity of the brood spots.

Often, birds also seem to sit with legs tied in knots. The swan generally sticks her legs behind her, and stilt-legged birds (like curlews and godwits) place theirs far forward. Terns often seem to practically stand over their nests. Like people, birds have to struggle with problems well ahead of the time their young arrive.

Birds that don't brood, like our familiar cuckoo, don't have brood spots. That's easy to understand!

There are several birds—ducks, for example—that pluck out the feathers over their brood spots. A possible explanation is that this presses the brood spots into better, closer contact with the eggs. However, some brooding birds take this trait so far that they practically pick the spots bare, sometimes causing bloody wounds.

Length of brooding varies considerably. There seems to be some correlation with size of eggs in some cases, but certainly not in all cases. There seems to be more of a connection with the stage of development in which the chick leaves the egg. There is a range, represented on the low side by well-developed "nest leavers", like chickens and ducks, and on the high side by poorly developed "nest huddlers", like house finches and canaries. Let us look at some specific examples.

American and European birds with short brooding periods are represented by the Chaffinch (*Fringilla coelebs*) and the American Robin (*Turdus migratorius*), which take about 13 days to hatch their eggs, with 11 days among the recorded minimums. In mid-range among brooders are pheasants (23 to 25 days), a Ring-necked Duck (24 to 27 days), and an American Eider (*Somateria mollissima dresseri*), which takes 26 to 28 days. At the upper end we have 27 to 30 days for the Kestrel (*Falco tinnunculus*); 28 to 31 days for the Ferruginous Roughleg (*Buteo regalis*); and 33 to 37 days for the Cooper's Hawk (*Accipiter cooperii*).

When I say, "The Cooper's Hawk broods 33 to 37 days," I mean that in the same nest, the hatch of the first and the last egg can truly occur about five days apart. That's a single clutch from a certain bird. But for the species as a whole, it is hard to come up with precise statistics because it depends on too many factors—heat, humidity, wind, hours of sunlight, night temperature, and other environmental conditions. Also of importance is the individual nature of the brooding bird, like intensity of brooding, frequency of switch with a partner, tightness of contact with the brood spot, nature and thickness of feathers, nest quality, and tempo with which eggs are laid.

Some people may not be able to accept the notion of this five-day spread because they believe that birds just don't begin to brood until their clutch is complete. Indeed, most clutches do hatch more or less simultaneously, especially with small song birds that lay eggs at a fast clip, normally one per day. Still, exceptions occur, in which within a small clutch of eggs there still is a full day's spread in the time of hatch. Birds that lay larger clutches very definitely don't hatch their eggs all at the same time. That's the rule with large predators, which lay eggs every *second* day. It's easy to confirm this finding on your own, if only by noting the clear difference in size among the hatchlings.

Which of the partners does the brooding—male, female, or both? For an easy rule of thumb, we can look at coloration of the parents. If they are more or less the same color, they are likely to take turns brooding. If the female is measurably plainer in plumage, then she generally broods alone. You never, or almost never, encounter a species where the *male* has a plainer plumage. The plain apperance camouflages and protects the female brooder on her nest (another example of mimicry). The flashy colors of the male serve as decoy, the male leading potential enemies away from the nest.

Does the male feed the brooding female? My observations confirm that he does. It makes good sense in species where the female is the sole brooder. The male often combines this service with another—guarding the nest and aggressively chasing away every trespasser. I am intrigued and impressed by this appealing show of cooperation between the pair. The falcon exemplifies this. While the female broods the eggs, the male hunts food, but he doesn't bring it right to her. She has to come and get it. When he has killed a prey, he settles down in the vicinity of the nest and calls her with several hoarse cries. She joins him and eats away from the nest to protect her eggs.

Hatchlings

Now, a few remarks about new hatchlings. With its egg tooth, the young bird breaks through the shell and pushes out with its body. All birds, sparrows as well as ostriches, have their heads near the air chamber of the egg while they incubate, so that all birds leave the egg from the same spot. The adults generally eat the remnants of the shell, or—if there are big pieces—the birds dump them away from the nest. That's one less attraction for preying visitors.

There are two basic types of hatchlings, which I call "nest leavers" and "nest huddlers"—terms which I mentioned in passing when discussing length of brooding. Young of the "leavers" (chickens, ducks, quail) dry off within several hours of hatching. Their down is heavy enough to protect them considerably. After just a few hours, they try out their legs and soon they are ready to have a look at the world beyond the confines of mother's feathery cover. On their second day, they are quite ready to join mother on a food-gathering jaunt, and they can peck up the morsels they find on their own. Adult care principally consists of leading the young to food and furnishing them cover under protective wings.

The "nest huddlers"—Canaries, House Finches, and birds of similar habits—present a quite different picture. The new hatchling lies exhausted in the nest, panting, powerless, and blind. Whether it dries off is

Olive Finch (*Tiaris olivacea*); 4½ inches (11.5 cm). Habitat: Cuba, Jamaica, and Haiti. The female resembles the male, but is duller, and has no black about the face. The chest only is mottled with black. *Photo by Author*

African Silverbill *(Lonchura or Euodice malabarica cantans)*; approx. 4-4½ inches (10-11.5 cm). Habitat: Senegal across to east Africa. They lay a clutch of 3-5 eggs in a little free nest or in a deserted weaver nest. The sexes are colored alike. *Photo by Author*

immaterial, because it is naked, except for a few scarce spots of thin down. The only positive thing it has is appetite, an insatiable hunger. Its food requirements really are enormous. Adult birds are relatively big eaters in their own right. Their activity level is high, and their body temperature is set high. A lot of food, readily digestible, must be available to keep their engines running. But young birds eat a great deal more. If they are to develop on schedule and in good health, they can easily eat more than their own body weight in a 24-hour period., The adults, therefore, are kept incredibly busy bringing a brood the incredible amount of food necessary for health and growth.

Ornithologists who like numbers have counted feeding flight repeatedly for a number of species. The count requires observation from sunrise to sunset because the trips to food source and back differ in length. The observer must record actual flights, not just project an average from a few samples. Nonetheless, I have done this work myself and became convinced that the statistics recorded by others are not exaggerated.

Dr. A. A. Allen, a highly respected professor of ornithology at Cornell University, made a motion picture of a female wren who made 1,217 food-bearing flights between a single sunrise and sunset. Dr. J. P. Thysse, a Dutch ornithologist, scored an even higher record for a Black-capped Chickadee, an amazing 1,356 trips.

Is this largesse shared fairly by the young? If so, how?

Nature helps bird parents wondrously in this potentially tricky job. All the adult has to do is pop the feed it's carrying into any random open bill. A helpful reflex takes care of proper distribution.

The system works as follows: When food reaches the throat of a young bird, a swallowing reflex is triggered, which works more promptly and more actively when the crop is emptier. Conversely, when the crop is fuller, the swallowing reflex is slower.

So, once a parent bird pops a morsel into an open mouth, the adult watches what happens. If the food doesn't get swallowed fast enough, the adult deftly removes the morsel and gives it to another of the young. Only the chick that swallows promptly and quickly gets to keep the food.

In addition to carrying food in, the adult also has to carry the young birds' droppings out. After bringing in a mouthful of food and watching it disappear, the adult pauses watchfully for a moment, observing the previously-fed young. When one rises to put its droppings on the nest edge, the parent bird snatches up the droppings. These are conveniently packaged by Nature in a type of sac, easy to pick up and drop off at a proper distance from the nest.

Practically all birds do this. It keeps the nest clean and keeps enemies away who would be attracted by the smell. However, one can still find some droppings on and near a nest whenever foraging trips take extra long. The digestive process then is too fast for the adults to keep up with.

26

Most small bird species have young in the nest only eight to 14 days. Generally speaking, the bigger the bird, the longer the nestlings stay.

Hatchlings are covered with down, not feathers, at the outset. The feathers develop between the down, beginning as closed, rolled up "stubble." At about a quarter of the length, the "flag" of the feathers then slowly rolls out.

Did you ever notice that a young bird is bigger and heavier than an adult? It is generally supposed that the development of an animal always progresses toward greater height and weight as it grows to adulthood. But that's not true for birds.

A young bird on its maiden flight can be 25% larger than an adult. It has just passed through an inactive stage with an excellent supply of food. Now enters a far more active stage with less food available and far more effort required to secure it. So the extra quarter of body weight is a food reserve to tide the youngster over a difficult time. A bird hatched in spring reaches its "normal" adult weight only in October or November.

That's pretty much our story about nests, eggs, and young. But the story isn't static. Look around for yourself in Nature and in your aviary. You will often discover something special. Not far from my home in the Netherlands, we found a Redstart nest in 1982 with a cuckoo egg in it. [The Redstart (*Phoenicurus phoenicurus*) is named not for its takeoff, but for its *tail*—"staart" in Dutch.] We also discovered a nest of Spotted Flycatchers (*Muscicapa striata*) that had the interior lined with toffee wrappers. And we found a lining of threads from coats and shawls in the nest of a Chaffinch (*Fringilla coelebs*). Take the opportunity for an outing and enjoy the novelties of Nature whenever possible.

Tri-colored Munia or Mannikin (*Lonchura malacea*); (11.5 cm). Habitat: central and southern India. The sexes are similar.
Photo by Author

2

Natural Bird Breeding

S UCCESSFUL HATCHINGS require a calm, stable environment. Don't start breeding too early. The safest period is from the end of April or early May until mid-September. Ordinarily birds breed three times a season. More frequent breeding should be discouraged to avoid risking the loss of females from egg binding.

As soon as the young are independent, they ordinarily should be separated from their parents. Otherwise, further breeding by the adults is inhibited.

In addition to seed, supply grit, or cuttlebone and charcoal. Adding several drops of cod liver oil to water or seed can be advantageous, but don't overdo it. At the most, use two or three drops for every two pounds of seed or quart of water.

Many hobbyists believe that seed-eating birds also raise their young exclusively on seed. That certainly isn't so. Most species feed insects. So, at brooding time, make available a steady source of ant eggs, small mealworms, fruit flies, aphids, grasshoppers, spiders, white worms and the like. Also provide old white bread, soaked in milk, twice a week, as much as the birds use up in a day.

Selecting and Managing Breeding Stock

Select birds for breeding that are not too old and not too young. That's important. For bigger birds, look up specific references giving the best time for breeding. Small exotics should be at least 12 months old—although there are species where the minimum age is eight months. Birds bred too young tend to become egg bound far too often, or they don't raise their young properly.

Smaller birds over four or five years of age are too old for breeding. The larger hookbills are not ready to breed before the second or third year and continue to do so for many years. Ordinarily, older males become sterile, or egg laying becomes difficult for the female—which also can cause egg binding.

Many cage and aviary birds have to be deliberately held back from breeding too early. Left to their own devices, they'd be sitting on eggs already in the heart of winter and would hatch young in that barren season. This early breeding drains the females, something that can cause unpleasant results in later hatchings. Young from the early hatchings generally are weak, which comes to the fore especially when they, in turn, are ready to be bred. And the breeder, also, should refrain from using sunlamps and other artificial gadgets to extend the breeding season.

The Breeding Season

So until the proper breeding season starts, keep breeding pairs separated—males with males and females with females. If you put on the right leg bands (of which, more later), sexing is easy.

If your birds start breeding at the end of March or in early April, you can expect two or three hatchings from most species. You will be able to distinguish birds hatched this way by their pure color and their healthy appearance—provided, of course, they were properly raised.

Be guided by the weather for when to start breeding. Bad weather can ruin a hatch—certainly in places where wet, chilly weather is normal in early Spring. You may miss a potential hatch by waiting out the bad weather. But that is more than offset by the fact that the young you do hatch stand out in quality and good health, and that your breeders stay healthy and strong, ready for successful breeding the next season. Conversely, if you breed too early, the young tend to come out weak, don't feather out well, and tend to get fatally ill from just a few days of exposure to bad weather. And your breeders tend to get sick and strained from overwork.

I can't emphasize enough that breeding pairs must be disease-free. Believe me. Healthy young can be raised only out of totally healthy parents.

Pairing Off

When breeding several birds, separate the pairs from each other before the breeding season. I use small exhibition cages, where I isolate selected pairs for about two weeks. I don't furnish any nesting material in these cages, nor do I attach brooding boxes. The birds just sit and eat, as well they should because a reserve layer of fat put on now will stand them in good stead later. The purpose of this isolation is to get birds used to each other—to promote *pair bonding*. Pairs bonded this way almost always stay together the whole breeding season after they are released into the aviary or colony breeding cage. I can check on whether my selected matings work by keeping track of individuals by their color-coded leg bands.

When several pairs of the same species are housed together, they should be homozygous (purebred) for color. This precaution prevents problems when pairs put together by the breeder don't stay together, once they come into contact with other birds of the same species. Birds, like people, prefer to select their own mates.

It happens that two birds you have selected as mates don't make it as a pair. You see evidence of this when a pair keeps starting and restarting its nest building; or, the female may lay abnormally large clutches of eggs, which the birds then cover up by a new nest several days later. I've noticed this especially in Zebra Finches.

The best response is to separate these unsuccessful couples and remate them with other birds. That seems to cure troubles of this sort in most cases.

Many aviary birds like to nest in a breeding box. Attach these boxes at varying heights, but not too low. Keep them away from perches and resting places to avoid birds soiling one another. Also, keep the boxes away from food and water containers.

Nest Building

Generally, tropical birds are industrious. They'll stay enthusiastic about raising brood after brood. This tendency is particularly strong in domesticated birds like Cutthroat Finches, Budgerigars, Canaries, Bengalese, and Zebra Finches, and shows especially in their nest-building activity. The males generally do most of the work, especially among Zebra Finches. They'll snatch up the nesting material immediately after it is furnished, and begin looking for a proper construction site, with a big piece of building material already in their beaks.

The male nervously goes about the site selection, while the female tends to stay calmer. She seems caught up in the important task awaiting her. She limits herself to ordering and straightening the nesting material brought in by her mate. There are exceptions, though. I have had females in my aviary that got involved with carrying material to a nest under construction. In other cases, I have seen a female who selected nesting material and placed her selections in the beak of her mate, for him to carry to the nest site. I like to observe my birds working together.

There are other birds, however, who are slow in starting construction. Don't force them, but rely on other pairs in the aviary to show them what to do. In most cases, slow nesters soon get the idea without need for action by the aviarist.

Be selective in the nesting material you furnish. I recommend hay and sisal-rope strands as basic material. If birds need anything further—like moss or twigs—look for specifics in the following sections about the special needs of different species. The two basic nest materials let the birds build neatly and come up with strong, safe nests.

Be sure to separate the rope strands and cut them into short pieces. I think pieces about 2½ inches (6 cm) long are about right. Longer pieces seem to be troublesome for birds to handle in nest building. They're also unwieldy for the birds to carry through the aviary, and tend to get entangled in their legs. Birds can get into serious trouble this way, and could be killed unless quickly rescued. The problem often occurs with quail and similar birds, particularly in species that have a strong growth of nails.

Furnish hay alone, at first. Then add rope strands later. The timing may be difficult in aviaries where several couples are nesting. But use your natural judgment and do your best to have breeding pairs keep close to the same schedule.

Don't panic, however, if breeding is not synchronized. If you show the birds clearly that building materials of two different types are furnished in separate racks, then they ordinarily will first use the hay to build the bowl of the nest and then the rope for the balance of construction.

Most species build rather coarse nests, but some build distinctly artful ones. Ordinarily, aviary birds use hay for the outside construction. The inside then is carefully covered with rope strands, horse hair, wool, feathers and whatever else might be available.

There are no absolute rules about nesting materials, although I counsel against randomly furnishing anything at hand. Wool, strips of toilet paper, hair, yarn and bits of textile material are all accepted. Small exotics, such as Zebra Finches, like carpet wool and they also welcome pieces of hemp rope. Natural colors are best for nesting materials. Birds will accept red, blue or green material—but with less enthusiasm.

It's a good practice to furnish only as much material as birds need for one round of nest building. Some birds can get carried away with the job, Zebra Finches in particular. They keep on building until there isn't a shred of nesting material left in the aviary. Even though the female does only the "interior decorating," she can get so absorbed in the job that she forgets about laying eggs!

When the nest is arranged to the birds' satisfaction—after three or four days of work—the female is ready for the amorous attentions of her mate. She starts beating her tail up and down and right and left as a sign she's ready to mate. The order of business—first nest building, then copulation—isn't absolute. Some birds mate first, then build a nest.

Mating usually occurs only on a single day. After several copulations—on a bush or twig, or in the nest—the female goes into the nest, ready to start laying.

Indian Silverbill *(Lonchura or Euodice malabarica)*; approx. 4½ inches (11.5 cm). Habitat: Afghanistan to Sri Lanka. The sexes look alike, but the male has a soft song and dances during the mating season. These birds make excellent foster parents, and have larger clutches (6-8/10 eggs) than African Silverbills.   *Photo by Author*

Diamond Sparrow *(Zonaeginthus guttatus)*; approx. 5 inches (12.7 cm). Habitat: south Queensland. The sexes look alike, but older hens tend to be smaller.   *Photo by Author*

33

Nest Checks

You may want to check on how the egg laying is going, but don't overdo it. It's wiser not to disturb a brooding bird, but there may be times when it's justified—even necessary. Don't worry if the egg count doesn't match the clutch size you read about in textbooks. Later clutches may be larger.

The reason nest checks are risky is that some birds react strongly to being disturbed while brooding. They may just abandon the nest. Most birds are not that way, fortunately. I remember having to move an entire nest, full of Zebra Finch eggs, from one aviary to another. The parents took the move calmly, continued brooding as before, and raised a first-rate brood of young.

You can check for unfertilized eggs by lifting them out of the nest with a plastic spoon. Don't use your fingers because the thin egg shell is damaged too easily. The smallest crack is fatal if untreated. Cracks, however, can be fixed. As gently as possible, mend the crack with scotch tape. It doesn't always work, but many an egg can be saved, especially a bigger egg.

When I check for unfertilized eggs, I do so after the female has been brooding eight days. Even that check isn't necessary. Most birds discard unfertilized eggs on their own, either by pushing them out of the nest or into the edge of it.

The best way to tell whether an egg is fertile is to *candle* it—to hold it up to a light. I have constructed a candling box with a compartment for the egg on top and a socket for a 40-watt bulb below. The egg compartment has a floor made of narrow-mesh woolen netting. I've seen ordinary hairnets used successfully in the same way. The important thing is to have a net soft enough to prevent cracking and a mesh narrow enough to keep the eggs from falling through.

A cursory look can distinguish fertilized from unfertilized eggs. The lamp light will show you the embryo in the fertile ones. You can see movement—the beginning of life!

Don't get carried away by watching the miracle of life because the intense heat of the light bulb isn't good for the embryo. Too much exposure would kill it. And don't candle the same egg more than once.

If you wait longer than eight days, you don't need a candler to check eggs. You don't even have to pick them up. Unfertilized eggs look pale red ("dirty red" some people say). Fertilized eggs, in contrast, can be distinguished starting about five days after onset of brooding, by their purplish brown, shiny tint.

After Hatching

All right! After waiting, patiently or impatiently, for "your" hatchlings, they have arrived. What do you do next?

Yellow-shouldered Whydah *(Euplectes macrocereus)*; 10 inches (25 cm). Habitat: Abyssinia or Eritrea to central Africa. The hen is similar to the cock, but has less black above.   *Photo by Author*

Nun Astrild or Black-crowned Waxbill *(Estrilda nonnula)*; approx. 4¼ inches (11 cm). Habitat: from Cameroon to Kenya. The hen is paler and browner gray above than the male.   *Photo by Author*

Interfere as little as possible. The parents know their task. They have preserved their natural instincts and follow them in the aviary much as in Nature.

So, control your enthusiasm. When you add to or replace food or water, let your motions be calm. If you show your hatchlings to visitors, have everybody talk quietly and without gestures. Keep visitors well away from the edge of the aviary. No matter how quiet visitors are, the birds get nervous just from having people around.

If all goes well, you'll soon see the hatchlings being stuffed full of food the parents have predigested in their crops. You'll see bits and pieces of the egg shells being eaten by the adult birds, while the larger shards are carried off or worked into the nest wall.

Depending on the species, you'll be able to hear the begging cry of the hatchlings after about a week. You may also note the fluorescent spots (papilla) on the beaks and tips of the tongues of the hatchlings. They differ somewhat in color and light intensity (depending on species), serving as a feeding guide and stimulant to the feed-gathering parents.

Mishaps can occur in the aviary as in the wild. One of the adult birds, on its way out of the nest, may get its leg entangled in one of the young and drag it away. The young bird falls to the floor, but if you pick it up quickly and replace it in the nest gently, it will probably recover surprisingly fast. You can reduce such accidents by disturbing the parents as little as possible.

Almost Grown

If an older nestling flutters out of the nest a few days too early, don't pick it up without thinking first. It may be oppressively hot in the crowded nest, where four husky youngsters may be jostling each other for space. If it is warm enough in the aviary, let the enterprising would-be fledgling spend a few days on the aviary floor. It will get the food it needs.

However, if the young bird is still small and poorly feathered, or if you expect rain or high winds, it is best to put the escapee back in the nest. Hold it in your hand in front of the nest opening and wait quietly till the birds inside calm down completely. Then put the bird back with its brothers and sisters.

It is quite possible that the same bird will leave home again the next day, often early in the morning. Then, just let it be. These precocious nest leavers are disquieting and tend to stimulate the rest of the brood to leave home, too. You don't need that! The basic rule is, it's okay for young birds to leave the nest when they're fully feathered.

Some birds at that stage still are more than pleased to keep accepting free meals from mom and dad. I enjoy seeing young Zebra Finches beg for food. They beat their wings passionately, almost dragging them along the floor, and they turn their heads in the strangest ways. If mom and dad don't pay immediate attention, the begging young start screaming for food louder and more piercingly until it's often possible to hear them several dozen yards away. After about two weeks, most young birds can pretty well find their own food, but even so, they still like to beg for it.

The older hatchlings don't confine their begging to their own parents, but extend it to other birds trying to feed their brood. The strange thing is that practically the whole aviary population helps raise the beggars and makes them independent.

Once hatchlings reach independence, you must keep them from breeding. Starting into egg production would be bad, because the young females are not yet fully developed internally. This is reflected on the outside also, since they also lack adult coloration. So remove the young birds, identify them individually and by sex with colored legbands, and isolate them in a separate cage or aviary: males with males and females with females.

Birds can be considered fully sexually mature when they're eight to 12 months old, although large species take longer. They could be used for breeding after that.

Closing the Breeding Season

With most birds, breeding ought to stop completely by September 1. Yes, one could start a new brood in September, but before the young birds are raised and independent, it's into October and November, generally not a favorable time for them. They turn out to be delicate, less attractive, and poor breeders the next season. They also tend to lay infertile eggs and abandon their nests—just to name a few problems.

Your energy is better directed at giving the aviary a fall cleanup and to taking a good inventory. Furthermore and more importantly—the birds will benefit from the nice, long rest. After all, they literally *worked their tails off* during the breeding season. Next year, they'll have to raise another couple or three broods. So, separate your breeding pairs and let them relax.

If you want to, you can cage a Canary, Pekin Robin *(Leiothrix Lutea)*, Green Singing Finch *(Serinus mozambicus)*, Budgerigar, parrot, or Mynah bird by itself.

The bird won't lose luster as long as it has human company every day. Most other birds need to be kept as pairs. It really depends on the species. In most cases, three pairs together are fine—but two pairs are definitely not recommended.

Mid-July is the time to protect Canaries from the danger of mosquitoes as they can be carriers of the much dreaded Canary pox, which is extremely contagious. *Photo by Author*

3
Breeding Canaries

THE DOMESTIC CANARY is a descendant of the wild Canary *(Serinus canaria)* and was introduced into Europe by Spanish colonists. When they conquered the Canary Islands in 1478, they immediately saw the commercial value of the local, happily-singing birds. First, the colonists only brought male Canaries home to Spain to sell. Then, they also brought females, so that their wives at home could breed some. But they took great care to sell only males to "foreigners."

The idea was that if females were not distributed abroad, Spain could maintain a monopoly on Canaries. Many males were sold to France, Italy, and Germany before the monopoly was finally broken.

The Universal Canary

The story goes that a Spanish ship was wrecked en route to Lisbon, and the cargo—which included Canaries—went overboard. Many of the Canaries are supposed to have reached the Isle of Elba off the Italian coast, where they began to breed naturally. Soon, the Italians began breeding from the stock on Elba, and sold pairs all over Europe, even to Russia.

In the middle of the 17th century, Germans started selling their famous Harz Roller Canaries. And around 1713, Hervieux published a list containing an amazing 20 color varieties of Canaries. Yet, that was only the start, and by the early 19th century, the Canary hobby began to be increasingly popular.

Different Types for Different Tastes

Canaries have been bred with emphasis on color, song, or body shape. Some Canary fanciers concentrate on color and are not too concerned about singing, while others care only about singing. At present, the color breeders are in the majority.

Already in 1555, Gessner wrote that the Canary was "the bird with the finest voice" (see *Historia Animalium* a "History of Animals"). Many breeders today still concentrate mainly on perfecting their birds' song and breed the so-called "singing canary."

Emphasis on color arose when breeders found that by crossing various strains, they could get a variety of coloration. The discoveries of Gregor Mendel were in vogue, and his theory of genetics was applied to create the most awesome and wonderful colors. In many cases, this led to a sacrifice of singing ability.

Still other specialty breeders are interested in body shape. Over the years, several different shapes have been developed in Canaries. The best known is the "Lizard," brought to England by the Huguenots in the 16th century. Some say the Lizard is entirely an English creation. The facts aren't clear, but, at any rate, the Lizard was perfected in England.

The Lizard gets its looks, a type of scaly outline reminiscent of a real lizard, from back feathers that are colorless at the ends until after the first molt. Older Lizard canaries get wider feather edges, which change the outline on their backs. This development is also reflected in wings and tail, which—like legs, beak, and toes—are supposed to be black. Today, there are also agate-brown and isabel Lizards. The London Fancy, once a popular variety, has now disappeared from the scene.

Canaries and the Beginning Breeder

Canaries are well-suited to the needs of a beginning breeder who wants to go beyond a pair of Canaries in a cage. To begin, you would do well to attend several bird shows to help decide which breed of Canary interests you most. The breeds I recommend to beginners are Borders, Glosters, and Rollers. These birds have a nice voice and could even be kept in an outdoor aviary.

In making these recommendations, I'm assuming a new breeder will not be showing his birds at the beginning. On that basis, it is advisable to acquire one male and two or three females at first, and eventually a fourth female. That's good for an aviary. If you bring in more males, they'll fight for the females. If you bring in more females, they, in turn, will waste energy quibbling over favored nesting places.

The beginner should have good results starting into Canaries with a new aviary or breeding cage. Canaries can even withstand rather severe winter weather as long as they have access to a space at night that is free of

Green Singing Finch (Serinus mozambicus); 5 inches (12.5 cm). Habitat: southeast Africa. The female resembles the male, but is duller, and has no black about the face. The chest only is mottled with black. *Photo by Author*

In April most canaries are either occupied with their nests or already have laid their eggs. It is advisable to remove the cock from the breeding cage after the first egg has been laid. *Photo by Author*

drafts and frost. It's best, however, to use unheated or lightly heated space indoors, in an attic if necessary. Be sure, at any rate, that enough natural light or good artificial lighting, like "Vita Light" (available from any good bird dealer or pet supply outlet) is available.

Proper Housing

Unless you're tentative about your new hobby, invest in a breeding cage (a crate/cage model) with several sections and a net in front. Netting can be bought commercially in a variety of sizes. In each section, you can place a breeding pair.

Put the breeding cage in an unheated room, outdoors. Don't place the cage on the floor; raise it at least 8″ (20 cm) off the floor. You can use the space underneath for storage.

The cage should be 24″ to 26″ (60 to 65 cm) long, 20″ (50 cm) wide, and 20″ (50 cm) high. This cage can be separated into two compartments—one for the breeding pair and its offspring and the other for separating the male if he becomes troublesome. In any case, put the young in a separate compartment once they are independent, or, shall we say, no longer dependent on their mother for food. If possible, separate mother and hatchlings only with a net, so she can pass food through the mesh to her little ones, if she still wants to.

When the Canaries aren't breeding, you can remove the separation, giving the pairs more room to fly around in. Ordinarily, that would be during fall and winter. The extra room keeps birds healthy and helps them return to breeding condition. Exercise is especially important for females, because just before the breeding season they tend to get fat, which predisposes them to egg binding.

Basically, Canaries are pretty sturdy birds. True, they are susceptible to drafts and humidity, but so are most other birds. If you take a Canary out of a warm room and put it into the cold, you can count on the bird catching a cold and dying a speedy death. Therefore, breeding cages, exhibit cages, or any other containers for birds absolutely must be kept away from windows. The temperature variations near a window are too extreme, causing a constant feather loss in the birds. This leads to colds and other ills. Find a suitable place for a cage and then don't move the cage around. Even in summer, my bird cages remain in their usual places. Moving the cages of house birds outdoors or near an open window on a sunny day is potentially quite harmful.

Some bird fanciers like to gather all their birds together in a single indoor flight cage. Others prefer using small cages for the males and breeding cages for the females (the type of cage not made up of equal sections). Only very few aviculturists put their birds in an outside aviary— at least not in locations with cold or wet winters. Many beginners like to put

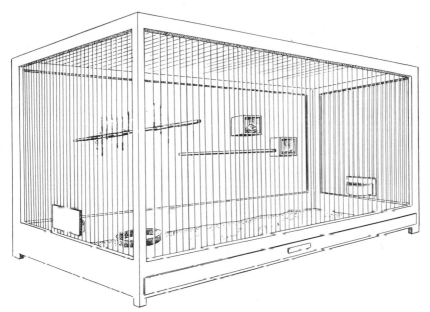

A large flight cage or indoor aviary. Excellent for finches, a few pair of love birds or a few selected soft-billed birds.

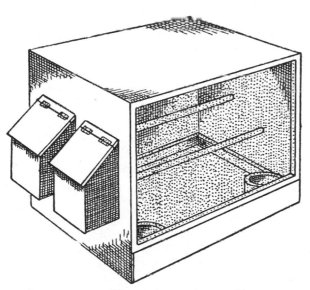

There are many different sizes and types of box cages available, but when the birds are to be bred, many aviculturists prefer this cage to all other types!

Drawings by Author

43

birds together, and there is no reason whatever not to set up community housing. Just keep the number of occupants low at first—as previously recommended.

Food and Feeding Methods

Commercial mixes are excellent for Canaries, and there are packaged seed mixes made for adults and young birds. They pose a disadvantage, however, in that some manufacturers tend to put too much rape seed and canary seed into their mix. In addition to these two seeds, the birds also need hemp seed, niger seed, linseed and some mawseed. Hemp seed is sweet and, therefore, a favorite of all seed-eating birds. It is also rich in oil, so give birds only a mouthful at a time, or they will get too fat. Niger seed also is rich in oil—remember that at breeding time. Consider that canaries which are housed outdoors should have more oil-rich seed than birds housed indoors.

In addition to seed, canaries should get grit, clean water, cuttlebone, free-choice green food (not frozen or too wet), sprouted seed, small insects, and a slice of old bread (soaked in water or milk). At the proper times, they also should get egg food. High-quality commercial products like rearing and conditioning food and a nestling food, also available commercially, are good for building up breeders.

If birds are accustomed to this diet, you can get them ready for breeding as early as April. This presupposes that you have indoor breeding facilities. If you breed outdoors with, say, several females and one male, wait until May if springs are cold and damp in your area.

Pair only birds in good breeding condition. The males sing loudly and the females are restless, flap their wings, cry for a male, and start carrying around pieces of green food, grass, or similar items—obvious preparations for nest building. If a male is aggressive, keep an eye on him. If he keeps on being troublesome after being paired, separate him from the female.

For nesting or breeding boxes, I use cubes 6″ (15 cm) on each side; however, there is no objection to other commercially-made breeding boxes. These commercial boxes can be fitted with rope nests, also available in stores. Before installing the rope nests, sprinkle a louse control powder in the boxes; be sure to get preparations labeled safe for birds and follow label directions scrupulously.

Assure good ventilation. If you use plastic, that can be a problem. Make a few holes in the plastic—but carefully, because plastic tears easily.

Furnish short pieces of unravelled rope, clean dog hair, moss, small roots, feathers, and the like for nesting material. Supply only a very small amount until the birds show that they are serious about starting construction and really begin work.

Canaries usually lay clutches of four to five pale blue eggs. The first egg is laid eight to 12 days after the female is paired, and brooding takes about 14 days.

44

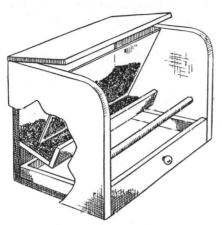

Seed feeder for mixed seeds.

Seed feeder in which the seeds can be offered separately.

Drinking vessel.

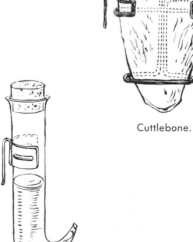

Cuttlebone.

Drinking vessel for hummingbirds and other nectar-feeders.

Drawings by Author

45

Most canary fanciers remove the first three or four eggs in a clutch and substitute fake eggs, available commercially. They want to be sure that all eggs are hatched at the same time, since most female canaries start brooding as soon as they have laid their first egg. During brooding, the male feeds the female on the nest. At that time, furnish rape seed, canary seed and egg food (rearing and conditioning food). Supply this last item only in the late afternoon. For green food, chickweed, dandelions cut into pieces, and spinach are all good.

Food for the hatchlings is a more difficult question. There is considerable difference of opinion on the subject and there are several types of hatchling food on the market. The quality commercial mix that works well for you is the right one for you to use.

Provide all commercial mixes in a crumbly form, but not completely dry. Use carrot juice to moisten the food.

You can also make kyour own food if you like. Start with three parts rusk, toast, or dry bread mixed with one hard-boiled egg. Make the mix crumbly with carrot juice, vegetable juice or skim milk.

Always have fresh food on hand. Don't supply more than birds use up in several hours. Be sure that the female doesn't overfeed her young. Symptoms of overfeeding are moister, thinner droppings that make the hatchlings look wet and dirty. When this state of affairs reaches bottom, young birds refuse to eat and die. To counteract this problem, put epsom salts in the drinking water and furnish soaked seeds and chickweed.

Again, a high quality rearing and conditioning food plus a slice of white bread soaked in milk makes a good hatchling diet. For the first three days, soften the seed you furnish—either by sprouting or soaking a handful of rape seed or niger for 24 hours. Small quantities of rolled hemp replenished daily also add to a diet's value. With all of this furnish plenty of green food as well.

Don't add chickweed to the green food suddenly if you have steadily been feeding only lettuce and other leafy vegetables; add chickweed gradually and in small quantities. Also, don't suddenly stop feeding chickweed. If there's any chance that your supply might run out before the end of the season, stretch out whatever you have. I've known of cases where the mother canary refused to feed her young any further because the breeder suddenly ran out of chickweed and substituted lettuce.

At three weeks, young Canaries leave the nest. At that point, remove the young and their father. Whatever feeding the young still need, dad can take care of. As soon as the young eat independently, move them to separate quarters.

If everything is coming along normally, the female can start a new round of breeding. If you leave the young with her at this point, she might featherpick the hatchlings.

At that time, be sure that the young can easily digest their food.

Hatchling food and egg food have to stay on the menu. Also, be certain the food doesn't spoil; eating spoiled food can be devastating to hatchlings. Take the soft food away every evening.

Keep cages scrupulously clean. Keep sheets of sand-covered paper on the floor. Remove it layer by layer as it becomes soiled.

Get young birds used to gradually eating hard seeds, like canary seed, rape seed, and cracked, rolled hemp. Start them on hard seed no earlier than six weeks of age, after they start flying. Gradually remove the soaked seed, leaving just enough for an occasional treat.

Not all breeders agree that young Canaries should get green food. Many people withhold green food until the young go through their first molt. Other fanciers offer green food as soon as birds can feed themselves.

Often, young birds suddenly lose considerable weight just before the first molt. They look sleepy, sleep a lot, and seem listless. Birds with these symptoms should be put in a warm place. Feed them soaked or sprouted seeds—no hard ones—plus white bread soaked in milk. Adding a little mawseed may also do some good.

When birds are eight days old, you can put on their leg bands for identification. If you belong to a bird club, you can probably order the bands there. Birds entered in shows must wear bands.

Around July 1, stop rebreeding female canaries, mainly to avoid egg binding.

Young Canaries go into their first molt near the end of the summer or in early fall, depending on the time they were hatched. They then are eight to 10 weeks old. They change all their feathers, except those of their wings and tail, over a two-month period.

Don't handle molting young more than absolutely necessary. They need a lot of rest to complete their molt in good health.

At that time, provide a rich and varied diet and replace drinking water daily.

Fanciers who raise red Canaries often feed their birds additives that enrich the red in their feathers. These additives are available commercially. Color additives also are provided for Yorkshires, Norwiches, and Lizard Canaries to preserve their distinctive markings. These additives can be included starting the seventh to 10th week of life, and are added gradually to the diet along with root juices, spinach and other vegetable derivatives. Commercial additives are best mixed into the Canary rearing and conditioning food.

Dedication and determination are important qualities for anyone interested in breeding Budgerigars. Developing birds that combine physical perfection with superb color is an intriguing challenge and one that attracts thousands to this hobby around the world.

Photo by Author

An opaline cinnamon sky blue Budgerigar. Good color and markings, good head and excellent deportment are evident in this lovely bird.

Photo by Author

4

Breeding Budgerigars

BREEDING BUDGERIGARS, probably the most popular of all pet birds, requires a real sense of dedication. It is not hard, I know, to get young from a pair of Budgies, but producing young of good quality can be a challenge.

This chapter is meant to help you get first-rate hatchlings that grow into birds you can be proud of. That should be your objective as a true bird lover. If you succeed, you will be a force for the good of this species, which needs rapid improvement in quality.

The Breeding Season

The breeding season for Budgies is not the same for all breeders; it depends largely on facilities. Indoor breeders, for example, can already start breeding in mid-December—using spare rooms, the attic, the garage, or a specially constructed indoor space. On the other hand, breeders who exclusively use an outside aviary can't start before March or April, depending on the weather. Still, both indoor and outdoor breeders should stop breeding by the end of July or the first week in August at the latest.

If the birds still show tendencies to want to go another round, don't be tempted to go along. They need at least a three-month rest. Separate the sexes and house them apart in spacious quarters. It is best if the former partners can't see or hear one another while they are separated. That's hard if you have limited facilities. Still, you can minimize contact by placing mats or other sight barriers between them. After a while, the birds hardly respond to each other's call because of the many other distracting noises they hear.

At this time, it pays to have a good system of identification and recordkeeping. Color-coded rings can help you separate males from females and aid you in restoring former mates at the next breeding season. Knowing who is who in the aviary is an important step toward responsible breeding. Good records also are useful at the time you sell a bird you've bred. The new owner then can continue breeding based on a knowledge of the bird's background.

Even if you have removed the mates from a group of females, they still may go through the motions of breeding. So be sure to remove all nest boxes and breeding facilities and take away any eggs they still may be laying. Give birds that don't want to quit breeding some flax climbing ropes, extra branches for gnawing and climbing, calcium blocks to peck at, or anything else that diverts their attention and natural drive.

Three months of rest from the tiring activity of the breeding season for both sexes is a sensible practice. Don't take birds out to shows till they have recovered. Major shows are usually scheduled in October, November, and early December, with regional or local events falling even earlier. Your best birds, which you plan to show, need to be selected in advance and trained for several weeks so they will quietly allow judges to see all their beauty.

If you want to leave yourself time for showing, simple arithmetic tells you that there can be no more than about seven months of breeding. Limiting the breeding season to two rounds, at any rate, is wisest. This way you take no chance on overstraining the female, which pays dividends in her remaining active in raising healthy young.

Most fanciers think I'm too conservative, preferring three rounds per season. But I stand my ground on this point: Two breeding rounds per couple per year gives ample time to expand your stock of good birds, provided you go at it responsibly and unhurriedly. Make quality your goal—not quantity. You can produce just about as many first-rate birds as birds of lower quality.

Age of Breeding Stock

Don't start birds into breeding too young. Many fanciers start breeding birds that are hardly nine or 10 months old, but there are great disadvantages in this. Biologically, females are not full-grown at this age, and their bodies are not able to produce eggs without risk. Birds that start laying too young are too nervous and unsure, something often misinterpreted as youthful enthusiasm. They leave the nest for the slightest cause, lay infertile eggs, don't brood some eggs, and, in short, aren't really up to the job.

One can make an exception to this rule only if necessary to rapidly establish or propagate a special mutation. Under such circumstances nine-month old females can start breeding if they are closely supervised and limited in the number of eggs they are permitted to brood. Give the eggs to foster parents to raise, to save the young mothers from the tiring effort involved in brooding and raising their own young. But this exception is not meant to convert young females into egg-laying machines!

Ordinarily a breeding pair of Budgerigars should be at least 12 months old, although the male can be only 11 months, but definitely not younger. This point is important because many bird fanciers are over-eager and rush into breeding as soon as they think their birds are grown.

In fact, however, birds like Budgies that reach adulthood at about one year of age don't reach their optimal breeding condition until their second and third year. Only then is the bird's body truly full grown and fully developed, and the bird has learned the tricks of brooding and raising young.

If you keep good records, you can check these facts, established by research, with your own birds. You will notice that results with birds in their second and third years are substantially better than those with birds in their first year. After the third breeding season, the performance of the female especially tends to decline, while the male's strength clearly abates after his fifth or sixth year. Young from later breeding seasons clearly are of lower quality, and champion breeders prefer working only with two- or three-year old breeding pairs.

Proper Housing

Proper housing is of prime importance, next to proper feeding and lighting. If you want to concentrate on raising just a few really top quality birds, you should house specially selected breeding stock as couples in separate breeding cages or mini-aviaries. If you wish, you can put unrelated species of other birds in with them—perhaps one of the smaller Australian parakeets or their relatives.

Separate housing is the only way to properly plan, control and keep records on your birds. Watch each individual chick and note any observation on a separate index card or other record-keeping system.

The alternative to breeding individual pairs is communal or "colony" breeding. If a colony consists of several pairs of healthy birds with the same color mutation, colony breeding is not such a disadvantage. It is harder to provide proper care to all the young, but breeding is done more easily and quickly. Even birds that have been domesticated as long as Budgerigars retain the habits of their wild counterparts in Australia, which breed in colonies.

Well-constructed breeding cages for Budgies are available commercially. Models that measure at least 24 inches long (60 cm), 20 inches deep

An aviary consisting of three parts: a night shelter, a half-open section (with the top covered by a sheet of corrugated fiberglass or similar material) and an open area, called the "flight".

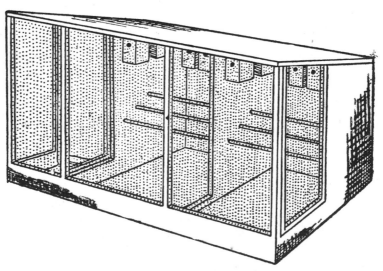

An aviary consisting of two parts; ideal for breeding parakeets, Budgerigars or love birds.

(50 cm) and 18 inches (45 cm) high will do nicely. You can build the basic cage yourself and buy trellised fronts on the market to fit the 24 by 18 inch dimensions of the home-built cage. You also can buy fronts 12 inches (30 cm) high with lengths of 24, 32, 40 and 48 inches (60, 80, 100, and 120 cm). I prefer the 32 inch length.

Cages require preparation. Even if bought new, they should be washed and disinfected thoroughly. Previously-used cages also are cleaned, submerged in water with disinfectant, and thoroughly dried out before use. If paint is needed, be sure to use lead-free paint.

Get good nesting boxes. Best for Budgies are the closed variety with an entrance hole about 1½ inches (4 cm) in diameter. The minimum inside dimensions are approximately 5 inches (12 cm) square at the floor and 10 inches (24 cm) high.

Cages differ as to access for the breeder. I like the type with a sliding bottom. There also are models with a door on the side or back—or a roof that swings open.

The floor of the nest box is completely filled with a nest base that has a gradually sloping bowl shape inside it with a diameter of approximately 5 inches (12 cm). In this nestbowl, put a handful of sawdust. Some birds will push this out of the box gradually, while others seem to appreciate it. Wild Budgies usually also prefer rotted wood and similar decayed vegetable matter in their nest holes.

You can attach the nest box to the inside or outside of the cage. I prefer the outside. Attach it thoroughly so it doesn't fall down by accident.

Get everything else ready before the breeding season. Complete your records. Check the lighting, heating, and thermostat. Buy your supplies.

Food and Feeding Methods

Don't forget to hang cuttle bone against the front of the cage near a roost. Stores also carry excellent calcium and iodine blocks for hookbills. Prepare a bin for grit and another for water. I also get a couple of stalks of spray millet to hang in the immediate vicinity of the nest opening, a trick I learned from English Budgie breeders.

The advantage of the millet stalks is that they draw the newly-introduced female to the nest box, so that she can become acquainted with it promptly. I refer to the female, because she gets introduced to the cage first.

Buy a starting supply of food. Be sure you have large enough seed bins, because birds should never be out of food. They should have enough each day to last till you come home. Remember that you also will soon have young to feed and they will need considerably more food. The youngsters will eat practically half their body weight in food during the work day—sometimes even more. In 24 hours, consumption can even exceed body

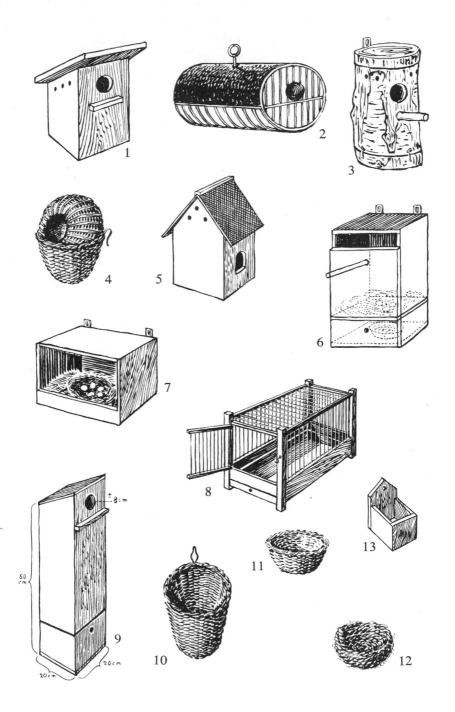

Nest Boxes for Every Need

Birds of varying species need nest boxes of varying design. The nest boxes shown here will suit virtually any small cage bird as follows:

1. Budgerigar, Pekin Robin, Zebra Finch
2. Australian Grassfinches, Mannikins, all small African finches, Zebra Finch
3. Budgerigar, Java Sparrow, Love Birds and the Zebra Finch
4. All species mentioned for #2 and the Pekin Robin
5. All species mentioned for #2, the Pekin Robin and Love Birds
6. All species mentioned for #2, the Pekin Robin, Java Sparrow and Love Birds
7. All species mentioned for #2, Canaries, Cardinals, large Finches, the Diamond Dove and Pekin Robin
8. Canaries, Zebra Finches and large Finch species
9. Parrots and Parakeets
10. All species mentioned for #2, Canaries and large Finches (The basket should be half-filled with nesting material.)
11. Canaries and large Finches
12. Canaries and large Finches
13. Canaries, large Finches and the Diamond Dove

Drawings by Author

weight. A female Budgie may make well over 800 trips to the food-tray, per day, to feed her young.

Do replace food daily, however, to keep it fresh. And get the parents used early-on to hatchling feed, so that they will give it to their young as soon as they are hatched. Also be sure to make fresh drinking water available daily.

Normally, Budgerigars get into breeding condition in November or December, the females somewhat earlier than the males. Naturally, both sexes should be in breeding condition at the same time, or else they won't mate. Note that the mating urge slowly wanes between the end of December and the end of January.

Continue to keep the sexes separated during this first phase of the mating urge. Give all birds an extensive menu of feed, including animal fats. Again, give them twigs of fruit trees, willow, or the like for diversion.

You can practically depend on the first breeding phase to end by late February, and in early to mid-March the second breeding phase will set in. By the end of March both sexes usually are in top breeding form. These dates, of course, are approximations, and experience will teach you when your own birds pass through the first breeding phase.

Daylight Triggers Breeding

The determining factor for the onset of the mating urge is the number of hours of daylight per day. This is true for wild and domesticated birds, both. Wild birds also are influenced by the presence of ripening grass, weeds, and grain and the availability of water. But these supplies don't materially influence the mating urge in captive birds, because the breeder keeps furnishing them all the time.

Artificial light affects the birds much the same as natural light, and this factor causes grief for breeders who aren't careful about regulating their lighting. If you want to trigger the mating urge, expose your birds to 13 hours of light and 11 hours of darkness. This means keeping lights on at night from dusk to 8 PM, and again in the morning from 7 AM till sunrise.

Breeders run into problems with day length when they get careless with their lighting schedule and leave lights on an hour or so longer one night and one or more shorter the next. This much irregularity is enough to upset the triggering mechanism in the birds.

The message in all this is that you have to be consistent in timing the lights. You can find timers that do the job automatically, some with dimmers or sophisticated modifications that allow you to make timing variations from three seconds to 30 minutes.

Expose your birds to as much natural light as possible. I don't leave the lights on beyond 8 PM in winter or furnish any lights in the morning. This makes the birds take advantage of the natural morning light, which

56

Nest Boxes for Every Need

Here are some additional designs which are acceptable as nesting sites to a wide number of species:

  1.-2. Budgerigar, Java Sparrow, Love Birds and Zebra Finches
  3.-14. Various Finch Species

11. This is a "3 in 1" model and is used for Budgerigars, Love Birds and Finches.

Drawings by Author

increases as the days go by. I do switch on a 15 watt nightlight for the birds, so they can get up and eat in the early morning if they wish.

A male in good breeding condition has a beautiful blue cere. He is in constant motion and takes in everything in his surroundings. He reacts to everything that interests him by pulling his feathers together tightly and carrying his wings high. And just about everything interests him!

A female ready to breed has a deep brown cere. She is also extremely active—especially in the morning. She constantly calls to her mate, and occasionally approaches him closely.

In forming or reforming breeding pairs put the female in the cage first and let her explore it alone for about 24 hours. Start about 11 AM, so the female will have time to get used to her new quarters. Usually she is taken from a larger enclosure to the breeding cage. I let my Budgies spend the winter in roomy flights, where their body systems can achive good condition. I prefer that to restricted caging in winter.

A newly-caged female sits uncomfortably in her new cage at first, which is not surprising, considering the change. She jumps nervously from roost to roost or hangs uneasily against the bars of the front of the cage. This behavior is most pronounced in females being mated for the first time.

I prefer to give the female immediate access to the nest box, but there is a difference of opinion on that subject. I think, however, that it is advantageous for females to experience the nest box as part of their new cage. You can check which method works best for you.

After 24 hours—again about 11 AM—the male can be put into the cage. You will notice right away that he gets accustomed to his new surroundings almost in no time. Often he reacts positively to the first approach by the female within the hour, but that can vary. It can take several days before mating begins.

At first there is no actual body contact. Both sexes, especially the female, gnaw at twigs almost constantly. They also gnaw at wire mesh and woodwork. In short, they always have something in their bills or fuss at something with their bills. All this should take place more or less in mid-March if you want to raise young from the first round of breeding to show in October or November of that same year.

Courtship and Mating

Mating starts with a type of little dance, consisting of a series of small dance steps along the roost, accompanied by both birds scraping their bills along the roost rod. Soon they rub their bills together and you can expect copulation soon. During the mating dance, and especially during the beak rubbing, the pupils of the eyes of the couple widen greatly.

But don't get close enough to check this for yourself. If you want to watch the mating scene, do it at a distance. I have found that successful copulation tends to take place in the early morning, between 6 and 8 AM approximately. It is an advantage to be sure that the desired mating has occurred—it can save you time and prevents uncertainty.

At a certain point, the upright position of the female changes into an almost horizontal position—parallel to the roost. Then she lifts her tail into the air. This is a sign for the male that she is ready to mate. He then climbs on her back, holding his head close to hers. He seems to wrap his tail around her rear section.

To maintain their precarious balance, both birds flap wildly with their wings, especially the male. He holds his cloaca against hers and with several rocking motions he inseminates the female and the mating act is complete.

Sperm is packed, as it were, in a semi-liquid, in which the sperm can move. The female organs also contain a liquid which is increased in quantity rapidly during mating and copulation. The sperm are thus able to swim up the oviduct to the ovum. Usually only one sperm pushes its way through the ovum's wall, although it occasionally happens that sperm fertilize more than one ovum with a first copulation. The ovum then forms a cyst, an impermeable wall that withstands the pressure of additional sperm trying to enter.

Sometimes the paired birds seem totally disinterested in each other. In that case, give the birds at least a week to get used to the situation. It may be that the birds just need extra time and opportunity to get used to their new surroundings. Females can be particularly fussy. Just moving a female from one aviary to another can make her feel ill at ease enough to act suspicious toward a mate with which she had previously been paired successfully. Similarly, females may not be used to a new owner and his work methods and habits.

If the birds still don't seem to want to get together, try to discover the reason for the rejection, usually by the female. It could be that you inadvertently put their cage in an uncomfortable location. It could also be that you're working with birds that are too young and you'll simply have to wait until they grow up.

With newly acquired birds, change in ownership could be a factor in poor breeding performance. The birds may have previously been bred under unfavorable conditions and now are reluctant to mate again. Or, it could be that a physiological problem is suppressing normal breeding activity.

Another reason birds refuse to breed could be a difference in day length. Not every fancier sets his timers the same, and it could be a previous owner gave eight hours of "night" while you are providing 12. You can tell if females fall out of breeding condition if their ceres lose their brown color.

Finally, the female may have previously selected a mate if she had been kept in a flight cage with males. So she can hardly feel attracted to another, suddenly-introduced male. She may absolutely refuse to accept him.

There is a fairly simple method to determine if two birds really aren't made for each other. Remove the male for a day or so and reintroduce him in the morning around 7 to 8 AM. If they don't get together then, you will have to find another partner for the female.

If all goes well, the female starts investigating the nest box and spends an increasing amount of time there. Sometimes she may spend the better part of a week in and around the nest box before she starts laying. At night, however, she is likely to sleep outside the box, though very near it.

Sometimes a female spends all night hanging from the bars of the wire mesh or the cage. It's a mistake to force her to "sleep right" on a roost or something apparently more suitable. The result of moving her would be to create a nervous and out-of-sorts bird, meaning a delay in further breeding activity.

After mating, the female no longer pays much attention to the male. She spends her time principally with inspecting the nest and its immediate surroundings. She keeps going in and out of the nest box and gnaws at the roof or entry of the box. To help dispel this nervous gnawing, provide her with several small branches from fruit trees, willow, or other soft wood. Put these in the nest, next to the bowl. She will continue to enjoy working on these once she is brooding eggs. It efficiently helps relieve her boredom.

Laying and Brooding

Seven days after copulation, the first egg is due, although it can take up to 19 days to arrive. Leave the female undisturbed at this time. She is very sensitive to disturbances.

Stop cleaning the cage for a while. Young females often get so upset from people cleaning their cages that they lay their next eggs on the floor of the breeding cage or stop laying altogether after producing one or two eggs.

Don't worry if the female pulls out several breast and stomach feathers. She is just preparing to brood her eggs better by baring her brood spots—places on her body where the surface temperature is higher than elsewhere because of an increased blood supply.

Young females often lay eggs on the cage floor without apparent reason. The eggs can be put into the nest box carefully if they aren't cracked or broken. The female may or may not accept such eggs for brooding—it's an open question. Most females, however, catch on quickly as to how things are supposed to go and eventually deposit their eggs properly in the nest box.

Budgies usually lay their eggs in the afternoon. After the female has laid the first egg or two, she tends to spend a lot of time in the nest box and she even spends the night there. Her cloaca relaxes and she is able to relieve herself again. When she leaves the nest from time to time, it usually is to defacate. Her droppings will be large and soft, compared to those of non-breeding Budgies.

60

Each additional egg produced is preceded by a new copulation, although it is quite normal that after the first copulation several additional ones take place the same day and the following days.

While laying eggs, the female may throw all the sawdust out of the nest box. I believe it is safer from the standpoint of preventing egg breakage to put the sawdust back—but do it very quickly at a time when the female is off her nest. If the female again exerts herself to remove the sawdust, then let it be.

The female will produce a new egg every other day. She may start brooding the moment she has laid her first egg, and if so, the first egg will hatch two days before the second egg and so on. If she doesn't keep brooding the first egg consistently, however, there can be a day's delay in the hatch. It even can happen that the first two eggs are hatched the same day. Ordinarily, however, if five eggs hatch, one can assume that the oldest young was hatched eight days before the youngest.

The first clutch consists of five or six eggs on the average. The second clutch can go up to 10 or 18. Especially in such large clutches, the infertile eggs should be removed. If you can't tell, or if there still are a large number of eggs left, distribute the extras among other breeding pairs of Budgies. Make certain to keep good records on the switch!

The incubation period lasts 17 to 21 days, depending on the temperature in the room and the relative humidity. In a warm room, with temperatures between 65°-72° (18°-22° C.) the embryos develop faster, but the brooding period never is shorter than 17 days

Normally one egg produces a single young, although twins do occur in exceptional cases. These twins, however, have a much lower chance of survival.

Brooding is almost totally the task of the female. At times, however, the male keeps his brooding mate company at the nest. They sit side by side on the eggs, the male with his head toward one side, the female with her head to the other side. But the most important task of the male is to take care of peace and security.

It is important for the breeder to know whether the eggs that are being produced are truly fertile. After three or four days you can usually tell the difference. If you hold a fertile egg against a strong light on a transparent plastic spoon, you will see dark red stripes on the upper yolk, a sign of new life. If the embryo is not visible after five days, the egg is infertile. But don't overdo the checking—once or twice is enough! You don't want to unnecessarily upset the birds.

Hatchlings

Eggs usually hatch in early morning. The hatching bird has an *egg tooth* on the upper mandible of the bill, which it uses to scratch its way out

A clutch of Budgie eggs benefits from the warmth of the feathers and other natural materials shown on the floor of this nest box.   *Photo by Author*

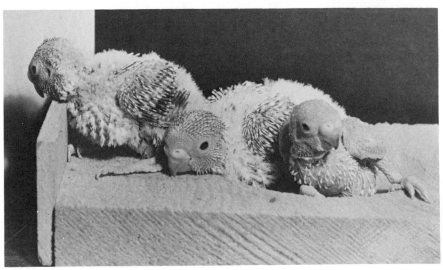

A trio of two-week-old Budgie babies.   *Photo by Author*

of the shell. Once it has wormed its way out, the little one will start softly begging for food, which the mother bird has made ready in her crop—the so-called *crop milk*.

Sometimes the male assists her, but ordinarily there is a definite separation of duties. The male fetches the food and offers it to his mate, and she, in turn, feeds the young.

The young are hatched naked and have pink skins. For the first few days, their protruding eyes are still closed.

Hatching doesn't always go without involvement by the breeder. Infertile eggs must be removed, because hatchlings would otherwise break them and make a mess in the nest.

You also need to decide once again if some females have too many hatchlings to feed. Distribute the extras in large broods to couples with smaller numbers of hatchlings. Select foster parents that have young of about the same age. Don't give either eggs or young over to the care of foster parents that didn't produce eggs or young of their own. They peck the eggs apart and kill the young. Don't delay in moving hatchlings around, as foster parents are more likely to accept young a few days old than halfgrown hatchlings 2½ to 3 weeks of age.

To promote acceptance of foster youngsters, especially older ones, rub their backs and wings with some droppings taken from the nest of the foster parents. They will then not give off their "own" odor, but rather the familiar smell of the foster parents. Use a small spatula to gather the needed droppings.

Mark eggs or young to be moved to foster parents with an odorless felt-tip pen and record the marking in your records. On live birds, put the marks on their backs. To avoid the possibility of mistakes altogether, give foster parents youngsters to adopt that have a totally different coloration. Don't depend only on feather color, but also use eye color. Lutino and cinnamon Budgies, for example, have red eyes and can thus be given to foster parents with normal eye color.

Banding

You can use leg bands on birds at least eight days old. Bird associations have leg bands available, and a diameter of 4 mm. inside measure is usually right for Budgies.

Putting leg bands on Budgies and other psittacines isn't difficult. They have four toes, two pointed forward and two pointed to the rear, forming a type of X. To start, wash your hands in warm water. Hold the young bird in your left palm if you're right-handed. (Left-handed persons do the reverse.) Lubricate the toes so that they adhere together and "squeeze" the three longest toes together between the thumb and index finger of your left hand so that they lie in a line with the ball of the foot. Take the band in your right hand, also between thumb and index finger, and ease it over the toes and the ball of the foot. The hindmost toe now is caught between the band and the leg. Take a sharpened wooden matchstick and shove it under the toe to liberate it. It's best to visit an experienced bird breeder and watch how he does it. If you have an agile, steady hand, the job is easy.

From Hatchlings to Fledglings

Young Budgerigars will show the first feathers on their backs and their entire bodies covered with down when they're about nine or 10 days old. At this age it's important to be sure they're getting enough food. This requires close monitoring which some people find a little tedious. See if the little crops are well filled and if the young birds are growing well and if they react sharply and alertly to your presence.

If the young birds don't have enough feed in their crops, you can feed them extra yourself. Use a feeding syringe, of which several models are available commercially. Or you can remove some underfed young for care by foster parents.

Females often will suddenly start laying new eggs while they still have young about 35 days old in the nest. Be sure to remove these eggs from the nest as quickly as possible if you want to save them. Otherwise the now already very active young will surely damage them.

Keep the eggs in a cool place, packed in a box with sterile cotton. Don't use dry sand for packing because it could harbor harmful bacteria. You can save the eggs up to nine days without risk, provided you turn them extremely carefully. Warm your fingertips before handling the eggs.

Alternatively, you could give prematurely laid eggs to foster parents.

The female may have such a strong drive to start brooding new eggs that she throws her old brood out of the nest. You'll notice that she is seldom to be seen in the aviary but spends long periods on the nest. And, of course, you'll notice the evicted hatchlings on the floor, loudly begging for food and trying to climb back into the nest with flapping wings and clawing feet. That's particularly true for older hatchlings.

You don't want all that commotion and the possible harm that could result. Take the young away from their mother and put them in a roomy cage with their father. He will take care of raising them further, with help as needed from you, the breeder. They can be housed without their father if they have started to peck and have clearly defined tails. House them indoors, if at all possible, because they really aren't prepared as yet to cope with night-time temperatures. Keep a close watch on these youngsters and feed them by hand if necessary to keep them fit. It would be a pity to lose them at this stage.

Under normal circumstances, young Budgies can eat on their own about a week after they become fledglings. At that point, separate them by sex and house them in a spacious run where they can develop further.

Pintailed Whydah, ♂, *(Vidua macrora).*   *Photo by author*

Senegal Combassou, ♂, *(Hypochera chalybeata amauropteryx).*
Photo by author

Yellow-faced Grassquit, ♂, *(Tiaris olivacea).*
Photo by author

African Silverbill *(Euodice malabarica cautans).*   Photo by author

Cordon Blue, ♂, *(Uraeginthus bengalus).*   Photo by author

*Left:*Chestnut-breasted Finch, ♂, *(Lonchura castaneothorax). Below:*Plum-capped Finch *(Aidemosyne modesta).* Photos by author

Indigo Bunting, ♂, (Passerina cyanea).   Photo by author

Above: Bengalese, or Society Finch, fawn, *(Lonchura striata domestica).* Below: Fire Finch *(Langonosticta senegala).*

A pair of the familiar and popular Cut-throat Finches, ♂♀, *(Amadina fasciata)* from East Africa.

Photo by author

Zebra Finch, ♂, *(Taeniopyg-
guttata)*. The model shown h-
displays the natural, or wild col-
of the species. Below: Red-crest-
Cardinal, ♂, *(Paroaria culculla-
Photo by aut-*

Management of Breeding Stock

To raise top quality young, in summary, limit the size of the brood. Personal preference and experience have led me to a limit of four eggs per nest. Next, limit the breeding season to two rounds. Otherwise you run the chance of having an egg bound female or young that are not properly raised.

Budgerigars being prepared for brooding must never be exposed to temperatures below 47° F (8° C). Take no chances on this score; you can't start breeding early in the year unless you breed indoors.

Good care and good management yield good results, generally, but there can still be complications. Eggs can be infertile, for a number of reasons. First, birds may not be in condition. Second, poorly shaped roosts may interfere with proper copulation. Round roosts are totally unsuited, unless flattened on top. Square ones are good. Swinging or loose roosts are also bad.

Third, be aware that there are females that jump into the nest box the moment they are placed into the cage, never to reappear, not even if a male is placed in the same cage. Further, check if the feathers around the anus are soft in both partners.

It's also important to remember that birds need proper artificial lighting and a constant supply of food and water to come into breeding condition. But, despite all else, birds need to have sufficient rest. That's why "lights out" at 8 PM is recommended. If lights stay on longer, the birds not only get tired but also quite nervous. It's best to dim light gradually, which is simple to do with the right type of dimmer switch and timer.

Note, finally, that there is a type of genetic defect that keeps affected birds from coming into breeding condition. Proper records can help trace such a problem. Successful matings can also be interfered with by a lack of vitamin E, more comprehensive dietary insufficiencies, nervousness, or under-age breeding partners. It's also important that the birds should always be able to hear each other's chatter. Interestingly, in quiet breeding quarters, where one hardly hears a sound, the rate of success in breeding is far from ideal.

Dealing with Eggs

Budgie eggs, according to measurements made in H. L. White's collection, range in size from 18.3 to 19.8 mm. by 13.7 to 14.7 mm., with an average size of 18.6 by 14.2 mm. This works out to approximately ¾" × ½".

Always handle eggs with the greatest care. The yolk and the developing embryo that is attached to it are held to the egg white by only two ligaments. A quick worker can easily turn an egg between his fingers without causing visible damage. But inside, all sorts of trouble can occur.

73

One or both ligaments could tear, so that once the egg is returned to the nest, the yolk gradually sinks through the egg white, settles to the bottom of the egg and attaches to the inside of the shell.

When the brooding bird turns this egg—as she will, several times per day—the misplaced yolk first gets too much and then too little warmth. As a result, the embryo stops developing and dies. The lesson in all this is to be extremely careful when handling eggs.

When checking on eggs, it's best to handle them with a plastic spoon. If you prefer using your hands, first warm your fingers. Cold fingers can kill embryos in their first stages of development, as strange as this may seem.

Remember that the egg you pick up should be returned to the exact same spot where it came from in the nest. If you aren't careful about this, the bird will return the egg for you, something that I've noticed repeatedly. This exposes the egg to more unnecessary movement, which is to the detriment of the embryo.

Check for possible dead embryos. Embryos can die at any stage, from early development until minutes before hatch. Still, the most common period for embryos to die is during their first week of development.

Between the 1st and 16th day of brooding, the most frequent cause of embryo mortality is a female that leaves the eggs too long because something has disturbed her. Such disturbances include sudden noises, a sudden, unexpected flash of bright light, and the appearance of mice, rats, cats, and other animals. Another cause may be a sick female, or one that has gone out of breeding condition—perhaps because the diet isn't quite right.

A proper diet includes food from animal sources, like egg food and universal food—commercial brands are fine. Don't just furnish these at the point where females begin brooding, but rather, start offering them several weeks earlier. Provide a choice of several brands to start with, so she can indicate her preference.

Embryos also can die if the egg shell gets heavily smeared with droppings. In my experience, a day or two after an egg gets dirty, the embryo dies. Smeared feces harden and form a seal, which clogs the pores in the shell, making exchange of gases impossible. Bacterial contamination compounds the problem, especially if the smears are caused by a female with diarrhea.

Other reasons for the deaths could be weak embryos produced by under age parents; poor housing, insufficient protection from cold and drafts, wet food (including green food).

Remember that embryo death is Nature's way of selecting out unfit birds. If you want to avoid embryo deaths, work only with superior parents. That way, you substitute your own selection for Nature's way—saving yourself unwanted losses.

Hatching

Hatching is obviously a critical period in a bird's life. Strong embryos from healthy parents usually hatch without much trouble. Embryos that are weak die, usually from lack of oxygen.

You can prevent the most common cause of mortality at hatching by making sure that the shell membrane stays supple. Chicks hampered by a dry, tough shell membrane have problems even if they hatch out. Chicks should be vital and active after hatching, not exhausted by the process.

Tough membranes that hamper or prevent hatch are caused by dry air in the birds' quarters. Central heating, especially, provides "dry heat." You can keep the membranes from drying out by putting several containers of water on the radiators or by setting up some commercially-made humidifiers. The best policy is to use both—you need to keep the humidity up!

It does happen that young can't hatch without our assistance. But don't be hasty. If you intervene too early, the embryo could die.

You timed your intervention right if an active chick emerges that has a pink-to-red colored body. You were too early if the new hatchling looks pale and white and a bit of the yolk is still attached to its belly. Such a chick, unfortunately, is doomed. And if you suspect the embryo is weak, don't bother to help it hatch. A bird too weak to hatch normally may live a day or two, but then will die anyway.

Let us suppose that you have an egg from which you suspect the chick can't liberate itself on its own. The 18 days of brooding have passed and the egg should be ready to open. You are expecting to hatch a totally healthy young—but nothing happens.

In such a case, I follow the approach of Budgie expert Gerald S. Binks, described in his fascinating book, *Best in Show* (Elbury Press/Pelham, London, 1977). He has prepared a table to guide your decision whether to assist or stand by:

SOUND	APPEARANCE	ACTION
Quiet tapping	1/16 in. crack	Too soon—replace
Quiet tapping	Group of fine cracks	Too soon—replace
Quiet tapping	Cracks plus brown line	Too soon—replace
Weak squeaks	Cracks plus tiny hole	Too soon—replace
Medium squeaks	Cracks and early discoloration	Too soon—replace

Loud squeaks	Crack line round circumference—creamy patches, moist membrane	Normal hatching—replace
Loud squeaks	Crack line round circumference—creamy patches, dried membrane	Assist
Loud squeaks	Large hole—drying membrane	Assist

Binks suggests laying the egg in question on a prewarmed, thick bath towel, with the part of the egg shell that has been cracked the most facing you. Cut a circle around the crack with a sharpened wooden matchstick. Be sure to cut through the membrane under the shell. You will see a little blood in the process, but if your timing was right and everything else went as planned, you should have assisted in bringing a healthy chick into the world.

If the chick you helped hatch is the first one to hatch, Binks suggests that you should be absolutely certain to put the broken shell back into the nest bowl. If you return a first hatchling to its mother without the shell, chances are great that she won't understand what has happened. She will regard the youngster as something foreign—a disturbance in the usual process—and she will kill the little one with a few chops from her beak.

There also are infections in the oviduct that interfere with peristalsis, the natural contractions that expel the egg. The remedy is to segregate the affected bird and to warm it up, as warmth can restart peristalsis.

Another threat to eggs is long toe nails on brooding birds, especially the females that do most of this task. Budgies can develop long, dagger-sharp nails that can pierce the egg shell and—if any eggs hatch—can hurt the hatchlings. If you notice any damaged eggs, remove them. There's only a slim chance they'd develop properly.

Cutting toe nails requires skill. Hold the bird up against the light and you'll see a red stripe through the nail, formed by blood capillaries. Cut up to the stripe, but have a coagulant handy in case you accidentally cut into the live part of the nail. See to it that the cut nail doesn't have a sharp point remaining.

Keep nails short naturally. Put some rough stones in your aviary or cage, or some rough-surfaced plants, like reeds or rushes. The birds will wear down their own nails in the course of normal activity.

Another problem to avoid with eggs are those with weak shells or no shells at all. Breeding birds need calcium in their diet to make strong shells.

Yes, some females lay shell-less eggs only, despite being properly nourished. It goes without saying that these birds should be retired from breeding. If you notice any shell-less eggs, remove them. They have no chance of developing and would break, smudging the normal eggs. As stated earlier, smudged eggs with clogged pores also fail to develop.

Oversize eggs are undesirable, too. They usually contain two yolks. Don't keep birds that persist in laying two-yolk eggs. Remove any such eggs you find unless you like to experiment to see what happens. The chances that you'll raise twins are minimal, but you could succeed! If you attempt it, keep good records.

A related problem to the double-yolk egg is the singleton chick. Some females lay only one egg per clutch, brood after brood. Such birds exist in greater numbers than one would suppose. Often there is no apparent cause. The birds come from a good line, are fed excellently, and can even score high in competition. Keep accurate records. If you are dealing with a heritable trait, don't breed from these birds.

If you do have to deal with a single egg in a nest, or if mishaps reduce the clutch to one egg or one hatchling, it makes no sense at all to have the mother go through brooding and/or feeding. It is extremely draining for parents to brood and feed their young. Research confirms that most of the demands and stresses on birds occur during the breeding season. I recommend taking the only chick and giving it for adoption to another nest with young of as nearly the same age as possible. Alternatively, if your schedule permits, you can raise the chick by hand.

Egg-Pecking

One of the worst occurrences in breeding is to encounter an egg-pecking bird. It often is hard to identify the culprit, certainly so if you breed in colonies or keep several breeding pairs in an aviary or cage.

A quick method I've found useful, especially when you have sudden evidence of egg pecking in the middle of the season, is to replace the eggs with fakes. You'll see which bird does the pecking, and the offender gets a good lesson! The real eggs that are intact are given to other birds to brood. Once again, keep good records.

If egg pecking happens in a one-pair cage, it is best to remove the male first. If he proves to be the culprit, you have no choice but to remove him after you have observed several matings and before you expect the first egg. You will have to do this after each breeding round if you continue to breed from him.

Suspect birds are not necessarily out of the running for the next round of breeding. Check if the diet is all right and provide extra vitamins, cuttlebone, and green food. Then put the affected cage as far away from others as possible for the next round.

If a female persists in pecking at eggs, you have two choices. First, don't let her breed any more for the season. Or, second, use a nest box designed to "harvest" eggs. These are oblong with a double bottom. The top floor is slanted, so that the freshly laid egg rolls away into a tray, lined with a thick layer of sand for protection. If you build the sloping floor yourself, it's all right to use plywood, because it doesn't stay in use long.

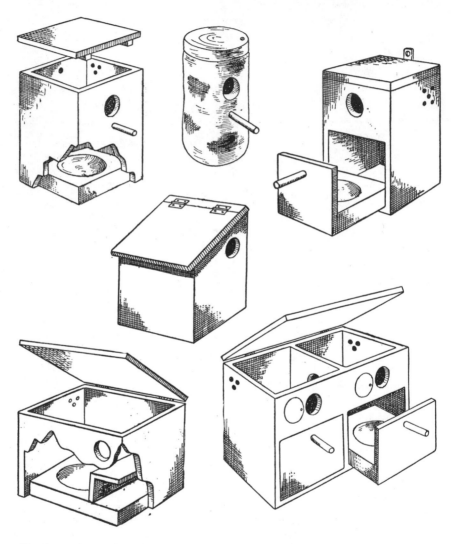

The six nest boxes shown here are used successfully in breeding Budgerigars and Love Birds.   *Drawings by Author*

Birds will stop pecking at eggs if they don't have any eggs around for a while and if you make sure their nutrition and housing are in order.

The same solution—a nest box with a double bottom—applies to another problem, namely birds that take their newly laid eggs out of the nest. They sometimes move the eggs in the oddest ways—none of which do the eggs any good. So remove eggs from birds that need to be broken of this bad habit as well.

Egg Binding

Another problem connected with egg production is egg binding. When housing and feeding are appropriate, egg binding shouldn't occur very often.

Egg binding is a condition in which the egg literally gets stuck or "bound" during the process of being laid. It strikes young birds in particular, but any bird can be affected when stressed by producing too many broods per season. That's why it's better not to experiment with trying to produce an extra brood after birds have raised their quota of young. Birds need a rest, even though it is possible to keep Budgies laying and brooding winter and summer.

Egg binding also can be caused by housing birds in flight cages that are too small, or by exposing them to repeated temperature changes.

You can recognize a bird suffering from egg binding by the way she usually sits on the floor like a little ball of feathers, shivering away. She lets herself be caught easily. Her rear end is swollen and her eyes have large pupils.

When egg binding occurs, you can relieve the immediate problem— the stuck egg—by helping it slip out. Smear some salad oil under the tail. Or else hold the bottom end (and *only* the bottom end) of the bird in a bath of alternately cold and warm water. Be sure the egg shell doesn't crack inside the bird, because that usually causes a fatal infection.

Birds that have suffered through egg binding shouldn't be rebred until four or five months have passed. Then watch them carefully to prevent a recurrence.

Cod liver oil, mixed into some seeds, can help prevent egg binding. Follow label directions to prevent overdosing the birds on vitamins.

Other Breeding and Rearing Problems

Other causes for problems in egg production are birds that are too fat to produce eggs without difficulty and danger. Or there can be cysts or tumors in the ovaries and/or the oviduct, which can cause the oviduct to rupture. Badly-infected birds like these are not suitable for further use in breeding. The same is true for males which develop infections in their reproductive organs.

The following remarks apply not only to Budgies, but also to other birds, including hookbills.

* Sometimes there are females, especially inexperienced ones, who overdo their job of brooding and sit on their eggs too long. One cause can be a temperature difference between the warm nest and the cool air outside. Obviously, people who want to start breeding the birds outdoors too early face this problem frequently.

The result generally is dead or injured young, since they are smothered by the mother immediately after hatching. They are too small and weak to draw her attention.

This analysis leads us to a solution. Take a young chick, about a week old, from another nest and put it with such a persistently brooding female. Smear the adoptee with droppings from the adoptive nest to get it accepted. Because of the bigger size of this foster chick, the female is forced to sit higher. The smaller, natural young can now move around, and soon are begging for food. This simple method usually works well—it has for me, and it is recommended by Binks and many others.

* Sometimes chicks have abnormally swollen crops, which prove, upon closer inspection, to be filled with air, not food. This problem points out the advantage of checking nests as soon as young are hatched—at the latest after a whole brood is hatched.

The solution seems cruel, but it is effective and painless. Sterilize a sewing needle in an open flame and pierce the crop with it, preferably at the place where the pressure is greatest. This allows the air to escape. The crop returns to its normal size and the chick is relieved of an unpleasant condition. If you don't relieve the bird, it wouldn't accept food and would starve to death. Continue to monitor the affected bird because the problem may recur and will require a second "operation."

* Sometimes a female doesn't start feeding immediately after the first chick is hatched. This usually is not the fault of the female, but of the young, which is too weak to raise its head, open its beak, and beg for food. The female won't start feeding unless properly stimulated: no begging, no food! The crying sometimes needs to be stronger than even a group of young can utter.

In such cases, again, an older "adoptee" can solve the problem with its loud, begging cries. You need to leave it with the new hatchlings only until the female starts feeding actively.

You can intervene directly with a little extra trouble. Remove one youngster from the nest and feed it a teaspoonful of baby food enriched with grape sugar heated to body temperature. You can obtain special plastic feeding syringes, which are frequently used by Canary breeders.

Now you have the task of getting the little beak to open. Lay the little one on its back between folds of a smooth, warm towel. It will immediately try to right itself. It that doesn't work—and it won't because the folds of the

A Budgie learns to speak quite
readily. A woman's voice, however, is the most suitable for
giving a bird speech lessons.
Photo by Author

Perches are placed at 45 degrees so that the Budgerigars on the higher perches do not
excrete on those below.
Photo by Author

towel restrain the bird—it begins to complain as loudly as it can. Then, put the feeding syringe into its throat and gently let some drops of food fall into the crop. Don't administer the food too forcefully to avoid getting it into the windpipe and lungs. (A mistake here would be fatal.)

After the crop is filled, clean the bird's head and beak with a flannel rag, so that no feed is left to become caked on the bird. Then, return the little one to its nest, where it will continue to complain loudly. This reaction will motivate the parents to start providing food on their own. It usually works with just one intervention by the breeder on a single young.

Of course, there are cases where a female won't start feeding, no matter what. Then all the young must go to foster mothers.

* Sometimes the parents stop feeding their young suddenly, after an unproblematic start. The causes can be many—among them cold temperatures, illness or serious disturbances. The only response is to end all breeding with the affected parents and distribute the young among other nests in the aviary. Remember to coat the young with feces from the intended foster nest—but don't overdo it. It is probably wise to give them a few hand feedings as described in the paragraph above. You can take on the entire job yourself, although it is preferable to use foster parents.

Feeding birds by hand ought to work well if the young are at least 10 to 12 days old. Be aware, however, that hand-raised birds become extremely attached to humans and very tame. Once they have their feathers and are independent, they can be taught all kinds of tricks. You can teach them to say a few words, even, and there are tapes and records on the market to help accomplish this. On the other hand, I have noticed that hand-raised birds are not suited for breeding, although there are rare exceptions.

* Sometimes one of the parents wildly attacks a young Budgerigar when it is leaving the nest. The victim generally dies. Binks (in the book named earlier) discusses this phenomenon.

The cause is not clear. In some cases I have examined the slain bird and have found that it was constitutionally weak.

The phenomenon may be related to that of wild Cuba Finches, which sometimes expel all female or all male young from the nest. This points to a natural population control measure, because research has shown that this happens in a year where there is a corresponding surplus of either male or female birds. I personally have seen wild Budgerigars kill their young and believe this could be another form of natural population control.

* Sometimes a parent bird dies during the breeding season. If the male dies, problems are relatively few, because a female usually can brood and raise a clutch of eggs and hatchlings on her own. If it is an exceptionally large clutch, you can distribute a few eggs to other nests. (Again, keep good records.) If the female proves unequal to the task, take all of her brood to other nests.

Lacking enough nests, ask for help from other breeders. You can see how important it is to belong to a bird club for this purpose alone.

If the female dies, procedures obviously change. If she dies on the nest, it may be hard to spot immediately. If you've been checking regularly, however, you'll find out soon enough.

You don't have to worry that the eggs are spoiled. Eggs often can take a lot more neglect than many people think, despite the remarks made earlier. A clutch that has not been brooded for several hours definitely can be saved. Put the eggs in a basket lined with cotton and place it under a lamp equipped with a 40 watt bulb for several hours, turning the eggs every hour. Check with your flat hand under the bulb—it needs to be comfortably warm.

Meanwhile, look for a way to place the eggs with other brooding pairs. Don't be surprised if these eggs eventually hatch a little later than the clutch with which they were put. There can be a two-day difference.

If the female dies while raising young, you can't expect the male to take over the job unless the young are about ready to leave the nest. I have known of several instances where the male has done an efficient job, but I still would keep an eye on such a brood to prevent accidents. If the young are older—at a stage where they are feathering out—you also can hand-raise them if you have the time. A feeding syringe, as described earlier, will be needed. In all other cases, distribute the young to other nests.

The breeder who has raised birds for some time knows exactly when something is going wrong with his stock. The birds eat little or nothing, sit with feathers fluffed out, and don't interact with each other. The breeder also can diagnose trouble from the looks of the nosecaps (ceres or wattles). Female Budgies that aren't properly housed and fed or bred too intensively have pale, almost white nosecaps. Stressed males have dark brown to black ones. The experienced breeder knows to react to trouble. If he has badly stressed birds, he will discontinue breeding from them and attempt to restore their health.

St. Helena Waxbill *(Estrilda astrild)*; approx. 4¼ inches (11 cm). Habitat: South Africa, Matabeleland, Madagascar, Mauritius, St. Helena and New Caledonia, with subspecies in southwest Africa, Cameroon, Loano, Nubia and Zambesi. The female is smaller and with a shorter tail and lighter markings. There is less pink on the abdomen.   *Photo by Author*

Blue-headed Waxbill (*Uraeginthus cyanocephalus*); approx. 4½-5 inches (11.5-12.7 cm). Habitat: east and central Africa. The female is similar in color to the Cordon Blue.
Photo by Author

5

Breeding Exotic Finches

IF YOU WANT to breed finches successfully, be aware at the outset that correctly sexing them is difficult. Many people have acquired what they thought was a breeding pair only to find out they really got two of the same sex. So if you're buying, be sure you get assurance in writing that you're entitled to an exchange in case what you got doesn't turn out to be what you wanted.

You might get some helpful hints from the coloration and shape of the birds in determining sex, but this is far from foolproof. The same is true in using song as a clue. True, in finches, only the males sing. But some restrict their song to breeding time only.

Breeding time, in other words, is the time of revelation. Song is one indicator of which birds are male. In addition, mature males ordinarily exhibit a characteristic pattern of mating behavior, evidenced by distinctive head motions, dances and displays of raised feathers. Young males, however, don't exhibit mating behavior.

Social Patterns

Social patterns and preferences of finches bear watching, if you want to breed successfully. If finches feel at all crowded, they fight. If there's a shortage of good nesting places, they fight. Males sometimes pursue females mercilessly. Even if just several ordinary pairs are kept under apparently favorable circumstances, they barely seem to be able to tolerate each other. Outside distractions also upset finches.

85

So, provide finches plenty of room. Don't overstock. Provide hiding places—like bushes and trees—in the aviary, so birds can get away from one another. Keep dogs and cats completely away. Limit visitors and your own visits to the aviary—the nests in particular. That's true especially at breeding time, when birds resent being watched.

Most domesticated birds are sensitive to disturbances. This includes Zebra Finches, Bengalese, Canaries, and parakeets. If you are using Bengalese as foster parents, they also should get plenty of privacy while in their parenting role.

When finches are disturbed, they tend to abandon the nest they're building and start a new one. They will also leave the nest if they're brooding, abandoning their eggs. Or else, they stop feeding their young or do a poor job of it.

The point is obvious. Finches need as much peace and quiet as possible, starting well ahead of the breeding season.

Feeding also requires special attention. Finches need a variety of food daily at set times. Put food containers at set locations some distance from brooding places.

Nests and Nesting

For brooding, you can use nest boxes, canary baskets, birch logs (with nest holes), closed or half-open boxes, and woven containers of heather, reed, and similar vegetable material. Make sure the birds have a choice of nesting places. Each pair should be able to choose between at least two types of boxes. To be more specific, if there are to be 10 pairs of finches in an aviary, provide at least 20 nesting places. Hang them at varying heights and locations, so the birds have a variety of choices. Keep notes on the preferences of different varieties of birds as a guide for the next breeding season. Keep in mind, however, that birds are individuals with their own "will," which is subject to change. A certain pair that selected a bush near their sleeping quarters one year won't necessarily choose the same location again next year. The chances that they will repeat, however, are good.

Give finches a wide choice of nest building material. Try sisal, coconut fibers, bits of straw, blades of grass, unravelled rope, tree bark, and dry moss. To finish off the inside, provide some wool (from animal as well as plant sources), moss, and down (from chickens or ducks). With finches, it's good to furnish nest building material at the very start of the breeding period. It stimulates their breeding drive.

Some birds prefer not to use nest boxes. For them, place wire mesh platforms or something similar in the bushes and trees of the aviary. You can also bring in nests abandoned by wild birds in your neighborhood.

Be sure to leave sufficient space between nesting places. If nests are close together, finches won't move in, at least not in general. If they do use

Gold-breasted Waxbill (Amadina subflava); 3½ inches (9 cm). Habitat: northern tropical Africa, south of the Sahara. The hen is duller, especially on the underparts.   *Photo by Author*

Quail finch *(Ortygospiza atricollis muelleri);* approx. 3½-4 inches (9-10.5 cm). Habitat: east Africa. The hen is paler and duller than the male.   *Photo by Author*

87

them, the situation leads to constant territorial battles between neighboring pairs.

Get birds used to a variety of foods long before the breeding season—insects (including some unusual treats), hatchling food, universal food (use several brands), bread soaked in milk and honey and water flavored with grape sugar. Also furnish weed seed and green food, extending through the brooding period.

Fledglings

Fledgling birds should be moved out of the aviary or breeding cage once their father has stopped feeding them for a few weeks. They create disturbances and prevent or disrupt the next breeding cycle.

Fledglings sometimes pull out each other's feathers because they are bored. To counteract this, give them weed seed, dandelions, chickweed, shepherd's purse, and plantain—a varied diet that should be available throughout the entire breeding season. You also can spread weed seed right on the floor. It helps stimulate the birds, relieves boredom, and motivates young birds to start foraging for food on their own.

Temperature and Humidity

Keep a close watch on temperature and humidity. Remember, exotic finches come from tropical or subtropical climates, where temperature and humidity are much higher than in temperate zones. In the tropics, the average temperature is 68° F (20° C), and the relative humidity ranges from 60 percent in places like central Australia to 90 percent in the mountains. You can try to acclimatize birds with special effort, but to breed them, you must approximate their native temperature and humidity. It's hard to do, especially with outdoor aviaries.

Electric heaters with built-in thermostats can help achieve the required temperature. In breeding cages, one can use an infra-red lamp with reflector, porcelain fitting, and clamps to hang the unit up. Don't use a bakelite fitting; it gets too hot, disintegrates, and eventually short-circuits. Electric appliances for birds kept in a cage or behind glass allow the breeder to maintain a constant temperature of 75° F (24° C), which I consider best for tropical finches normally bred by fanciers. It's essential to have a so-called "minimum-maximum" thermometer, one which indicates the highest as well as the lowest temperature in the area. There also are commercially produced combination instruments that measure both temperature and relative humidity.

If the outdoor humidity in your area is too low, you can spray an outside aviary with a garden hose every afternoon. Indoor humidity takes more trouble to maintain, and you'll need a hygrometer in the nesting area.

Pearl-headed Silverbill (*Lonchura caniceps*); 4½ inches (11.5 cm). Habitat: East Africa. The sexes are alike. *Photo by Author*

Bronze Mannikin(Lonchura or Spermestes c. cucullata); 3½ inches (9 cm). Habitat: the coast of west Africa across to the Sudan. The sexes are colored alike. *Photo by Author*

Orange Bishop or Orange Weaver (*Euplectes orix franciscana*);
approx. 5 inches (12 cm). Habitat: Senegal to the Sudan. The hen is
brownish, with dark streaks in the feathers.   *Photo by Author*

Parson Finch (Poephila cincta); approx. 4 inches (10.5 cm). Habitat: the eastern region of
northwestern Australia. The hen's head color is of a different shade, while her throat-patch is
narrower, and straighter at the sides than that of the male.   *Photo by Author*

A hygrometer needs to be primed before use. Wrap it in a *moist* towel, and keep it wrapped about 20 minutes; then the wire will indicate 95 percent. Adjust it with the set screw on the back. Clean the hygrometer four times per year; blow the dirt out—don't use water.

Proper humidity is very important to successful breeding. Too dry an atmosphere makes feathers look bad, and that isn't the worst of it. Egg membranes also dry out in the face of insufficient humidity, making them tough, and causing problems with hatchings. During brooding, eggs normally lose 14 percent of their starting weight through evaporation from the thousands of pores in the shell. This evaporation must be minimized. Personally, I like to keep the relative humidity in breeding cages and bird rooms, inside aviaries, or behind glass around 70 percent. It's not hard to achieve with containers of water, a spraying rig, or a specialized electrical humidifier.

Day length also is extremely important for successful breeding. Indoors, be sure that the hours of artificial lighting correspond to the hours of natural daylight outdoors.

Leg Banding

If you are going to enter your exotic finches in shows, they have to be leg-banded. It's hard to give a specific age at which you can start banding birds. As a rule, don't begin the job until baby birds start dropping their feces over the edge of the nest (or—in some cases—right on the edge of the nest). Before that stage, the parents remove all foreign objects from the nest (along with the feces), and thus, they would remove the leg-bands as well. The adult birds are so determined about the cleanup job that they literally throw out their young along with the leg-bands.

Generally, this stage is reached when birds are seven days old. At that point, they already have some small feathers, have well-developed wings, and have lost all or most of their down.

Leg-banding finches is harder than it seems. You need to stick the three forward toes together with petroleum jelly and press the rear toe against the leg. Then you can ease the band over the toes. Use a sharpened wooden match stick to flip the rear toe through the legband and hold it in place.

Be sure that cages and inside aviaries always are well ventilated, but not drafty. Keep floors dry, particularly when there are young birds present. You have less control, of course, over air movement in outside aviaries.

Finally, breeders of exotic finches should also read the section on Zebra Finches, which follows. The general remarks in the introduction will apply to all exotic finches.

Zebra Finch *(Taeniopygia guttata castanotis)*; approx. 4-4½ inches (10-11.5 cm). Habitat: Australia, mostly inland districts.

Photo by Author

Good exhibition Zebra finches are developed only through a strict, continuous, selective process and breeding with a great deal of care and attention to details.

Photo by Author

6

Breeding Zebra Finches

ZEBRA FINCHES certainly are among the most popular aviary birds today. They don't require a great deal of special care, they are excellent songsters, and they are easy to breed. They are exceptionally suited for the beginning breeder, and they still offer a challenge to the advanced breeder, who can pursue the challenge of breeding for different color variations. There's something for everybody in the Zebra Finch.

Zebra Finches have gone through a great deal of development, as many articles in bird magazines have discussed over the past few years. This development still continues.

People experiment enthusiastically to perfect color and body shape, something that can be achieved as well through crossing Zebra Finches with other species. It's a fact that breeders of Zebra Finches follow definite fads, and everyone concerned needs to adjust to this situation.

Classification

For a long time, naturalists have classified Zebra Finches along with many other so-called "weaver finches," under the subfamily *Estrildinae* of the true weavers, the *Ploceidae*. This classification has been reconsidered in the light of the major differences in anatomical structure, feather design, habits and ecology. (Ecology refers to relations between an organism and its environment.) Currently, Zebra Finches are classified in the family *Estrildidae* (see *A New Dictionary of Birds*).

The change in nomenclature seems logical especially in view of nest construction. True weavers always start building by weaving a few blades of grass or strong stems around a twig. Afterwards, they form a circular structure that serves as a side. Next, they build a roof. Only when they have completed the entire exterior, do weavers start building a brooding chamber.

Estrildidae handle the job quite differently. First, they build a dish-shaped construction in the fork of a tree. Then they build sides and roof. They never actually "weave." That's why the name "weaver finches" is confusing. It's better to consider them "grass finches," or *Erythurini,* especially because this group subsists mainly on grass seeds and builds nests from blades of grass.

Classifiers have had other struggles with the Zebra Finch—especially in deciding whether there are distinct subspecies. The scientific name for the Zebra Finch is *Taeniopygia guttata castanotis,* established by Gould in 1837. G.M. Mathews in the authoritative 12-part work, *The Birds of Australia* (London, 1910-1927) described separate Australian forms named *alexandrae* (1912), *hartogi* (1920), *mouki* (1912), *mungi* (1912), *roebucki* (1913), and *wayensis* (1912). In fact, these are all taxonomically identical to *castanotis* because in Australia there is but a single type of Zebra Finch, although there are color variants in certain areas. Mathews was not alone in supposing that there are three or more (up to seven!) subspecies.

Keast, however, definitively proved that in Australia no variants of the Zebra Finch exist. This may sound strange considering the enormous size of Australia and the yearly extreme drought. This forces the birds to move from Central Australia to better watered areas, which could easily lead to interbreeding with birds from different areas.

These comments should not be taken to mean that all Australian Zebra Finches look alike. Individuals do differ in size, beak color, and breast markings.

Outside Australia, moreover, there is a separate and officially recognized variant Zebra Finch. It is found on Timor and some of the other Sundi Islands, Letti, Sermatta, Luang, and the Moa Islands. The color of this variant is brownish yellow and the top and back are darker than those from Australia. One bird expert, Delacour, wants to include these birds, which he studied in Flores and Timor, in a single species with the Australian Zebra Finches. He wants to include them with the fancy finches *(Amandini)* instead of the grass finches. He "invented" the name *Poephila guttata castanotis* for all Zebra Finches—"Poephila," Greek for "grass lover"; guttata, Latin for "drop-like markings"; and "castanotis" for having chestnut ears. I disagree with Delacour, and most other modern references call the Zebra Finch *Taeniopygia guttata* (Veillot, 1817).

Feeding

The diet consists principally of ripe or semi-ripe grass seed and insects. Zebra Finches are known to hunt flying termites and other small, flying insects, as well as crawling insects found on the ground and on leaves. They often jump up from the ground to snatch grass seeds.

Zebra Finches are true seed eaters. Tropical seed mixes are easily available commercially, and the established brand names will give good results. Once you develop some expertise, you may well want to make up your own mix. But remember that it must contain the most important types of seeds.

Birds have a bad habit of treating themselves to seeds that taste very good but are not particularly good for their health.

In addition to the usual food, supply grass seed, loose or in bunches, plus spray millet. In addition, be sure birds get fresh water daily for drinking and bathing. In a sunny corner, spread some pure river sand. Birds absolutely require a sandbath to promote molting and fight off external parasites.

To keep birds in a truly first-rate condition, give them some extras. The first essential is green food, like chickweed, lettuce, endive, and spinach. Also, incorporate cuttlebone, which is made from the back plate of the common squid. (You can find it easily on the beach or buy it in a bird supply store.) Further, supply ant pupae and meal worms—small ones or big ones cut into pieces. These bring variety to the menu, a health-promoting factor for any living being.

Once the young are hatched, add the following additional items: rearing and conditioning food (be sure it doesn't spoil and renew it daily) plus white bread soaked in water or milk (preferably whole milk). Take care not to put the bread in the sun or it will turn sour. Every day, furnish fresh bread and remove old leftovers.

You can give birds their seeds in a self-feeder or in a mix. I personally prefer separate self-feeders for each type of seed. If any type of seed runs short, you can see this immediately and replenish it accordingly. The same is true with vitamins and other additives. Self-feeders allow birds to snack on just that feed which they need at the moment, without having to pick and choose out of a mix. This keeps them from throwing feed around, which is a real waste. And mixes only provide a certain percentage of each type of seed. Zebra Finches drink like common pigeons—they suck up their water rather than scooping it the way chickens do.

Calls

When the birds are happy and well fed, they call out with a trumpet-like "tee-tee-tee." When they fly in Nature in large flocks, they also give out the same call. If some get separated, they will call out with a much more piercing "tee-tee." This must be considered a sound of alarm because they also use it to warn of danger near the nest.

To maintain contact in flight, they call out a low "tet-tet," which is heard uninterruptedly in large flocks on the wing. It is a communication call. The same call is used for pair formation. Then the call sounds quite rapid, and it seems as if it is uttered in pain. A male Zebra Finch also uses this call to bring his mate's attention to a suitable nesting place, but then the sound is considerably higher in tone.

When Zebra Finches chase each other, one frequently hears an aggressive "wsst," which Immelmann has cleverly compared to "the tearing of a cotton rag."

The true song consists of low trilling and a loud "tee-tee" call. Each phase of the song is repeated several times, generally without variation. Only wild birds sometimes have some variation, which consists of the communication call just described.

Zebra Finches in confinement are well-known for a distinctive "trumpeting," which has not been heard in wild birds. Only the males occasionally sound off like this, and it seems as if it is a sort of expression of contentment.

Singing of wild birds in Nature can be heard most often when birds are alone and also right after pairing has taken place.

Pairing

Practically all grass finches do a mating dance during which they hold a blade of grass, a piece of straw, or a feather in the beak. Zebra Finches never do this. Regarding their mating dance, the ornithologist Immelmann says:

> During the opening phase, male and female jump up and down between two twigs on a branch, while they repeatedly rub their beaks. Their tails are oriented toward each other and go up and down. After considerable hopping up and down, the female stops and the male approaches her along the branch in a rhythmic, turning dance. With every jump he turns his legs and his body, singing constantly. His tail keeps on pointing toward the female and his crown feathers lie flat against his skull. The feathers on the back of his head and his cheek feathers are raised up, causing the black and white portions of the head to be accentuated, along with the chestnut color of the cheeks. Also the white stomach feathers are preened. All this is done to impress the female.
>
> As the male moves down the branch toward the female, her tail starts bobbing up and down and she begins to quiver. Next, the mating takes place.

In captivity, the mating dance is practically identical. The dance and the mating can be repeated several times. After mating, the male puts himself into an erect, horizontal position and quivers up and down with his tail—paralleling the behavior of the female. That's why D. Morris calls this "pseudo-female behavior."

Usually the mating dance takes place in dead trees or bushes because the leaves of living vegetation seem to limit the movements of the birds. Not infrequently, mating takes place on the ground or on rocks.

Ornithologists believe that in the wild, Zebra Finches mate for life. In captivity, however, the birds don't stay coupled nearly as much. In the wild, the couple stays together the entire year, even if they join a large flock. They live together in a comfortable sleeping nest to which they retire at sunset. At times, they also go into their nest during the day to protect themselves against the intense rays of the sun.

Nest Building, Eggs and Young

Zebra Finches usually nest in low, thorny bushes or small trees. Sometimes, however, they nest on the ground, in grass clumps, hollow trees, rabbit holes, and similar places. At times they try unusual sites, such as termite hills, the substructure of nests of predatory birds, nests of other grass finches (which are taken over without alteration), and swallow nests. Nests also are found under roofs, in gutters, and in holes in fence posts.

The nest is built from rough grass and is furnished on the inside with soft grass and fluff from fruits. The brooding chamber is lined with feathers, rabbit hair, sheep wool, and fluff.

In areas without grass, the birds use fine, thin twigs and roots for the external structure. Nests in bushes and other vegetation have been described as "bottle shaped" with the entrance on the side. If Zebra Finches build nests in a hollow, they sometimes don't make a separate roof.

Outside of breeding time, they often build a play or sleep nest. It differs from the brooding nest in that it usually doesn't have the "bottle neck," the little entrance tunnel. Yet at times the birds will use old brooding nests for sleeping.

Brooding nests are constructed by both partners. Generally the male concentrates on fetching building material, which is used by the female. However, that's not always the case. Immelmann says: "In Central Australia, I have noticed that both sexes carried building material." I have noticed the same thing in Southwest Australia and Queensland. I think both birds work together to finish the nest quickly to take advantage of the rainy season. Normally, the birds don't rush and sometimes it takes almost two weeks to complete a nest.

The breeding season differs in various parts of Australia, because a successful hatch depends principally on the rain. In Central Australia, Zebra Finches start nest building immediately when the first rains come, irrespective of the season. The reason is that when it rains, vegetation develops, assuring a food source for the birds. The rainy season can be quite short, so that the birds start building nests quickly to be able to raise at least one family before it gets too dry. In regions with more rain, like North Australia, the Australian "sparrows" breed from October to April.

The success of breeding in the wild also depends a great deal on the temperature. This is especially noticeable in Southwest Australia. The winter is too rainy to assure successful breeding. That's why the birds breed in fall and spring. In the summer, it's too dry and in the winter, it is much too cold. But in an occasional summer when it does rain, Zebra Finches promptly start breeding. In other regions, like East Australia, Zebra Finches breed all year, except in June, July, November, and December.

A good clutch consists of three to eight eggs, normally four to six. They are light blue to white in color. The size varies. Immelmann has recorded eggs with these dimensions: 9.8 × 13.8 mm. and 11.6 × 16.1 mm, which corresponds to an average of 10.7 × 14.95 mm.

Brooding begins after the fourth or fifth egg is laid and is performed by both parents. Male and female relieve each other in a way that avoids attracting potential enemies. It is worth noting that the male frequently carries a blade of grass, fluff, or feather in his beak. The female never does this, as far as I know. Most likely, he incorporates it into the nest edge—a form of relaxation. At dusk, both partners go into the nest, and by the first morning light, the male leaves.

The young are born after 12 to 16 days. There is no exact length of brooding time—too many variables affect this. For example, brooding intensity and weather conditions are involved.

Newly-hatched Zebra Finches are flesh-colored and almost completely featherless. Their skin darkens considerably at three days of age; at one week, they are black. Their beaks also start out light in color but turn black after a week.

The young are born blind, but open their eyes in eight to 10 days. At that point, the first feathers appear. For the first two or three days, the young are completely voiceless, but after that they make soft "sawing" sounds. These are the begging cries which become increasingly intense, loud, and shrill after a few days.

Both parents fetch food. They bring half-ripe seeds of grasses and other plants. They also bring soft, shell-less insects. After about three weeks, the young birds leave the nest. The timing depends on the available food, the weather, and other factors. In the beginning, the parents lead their young back to the nest in the evening, and they spend the night there. The parents also tempt their young back to the nest during the day by offering special tidbits from their beaks. However, this behavior is rather rare in captive birds.

Family ties are preserved a long time. In the evening, the family returns to the parental nest to sleep. The young start becoming independent when the parents begin building a new nest for the next hatch. Still, some

A wicker basket often makes a very satisfactory finch nest.

Donnelly

An assortment of man-made finch nests and some of the nesting materials finches will accept.

young regularly return to the parental nest, which is standing empty in any case.

The first molt begins after two months and lasts one to 1½ months. The beaks turn red and are fully colored after about 10 weeks, although they don't attain the fancy red of adult birds' beaks. Young bird beaks always are noticeably lighter.

Social Life

Zebra Finches are social by nature. Outside the breeding season, one can see huge flocks of wild birds, numbering from 40 to 100 or more individuals. If weather and food supply are favorable, the birds stay in the same area almost all year. They build sleeping nests there and couples retreat to them at night.

In areas that experience long droughts, like Central and North Australia, Zebra Finches migrate to areas where they can find water. There one often sees flocks of hundreds, even thousands of birds.

Zebra Finches like to live close together even during the breeding season. Colonies are relatively small, consisting generally of five to 30 pairs. Each couple generally has its own bush or little tree.

Where vegetation is scarce, however, two or three nests may be found in the same tree or solid bush. And this habit may expand. Prof. K. Immelmann quotes ornithologist Whitlock in *Australian Finches* as having found 13 occupied nests on one big hakea bush in the Wiluna district in 1948. McGilp is cited as having found 21 occupied nests in an acacia tree near Oodnadatta.

Immelmann believes that members of such a colony have closer contact with one another and can probably recognize each other by their call. Neighbors are allowed in each other's nests, even though strangers are aggressively resisted. Several times a day, members of a colony go to a certain spot to drink together, bathe, and groom each other. Sometimes the social drive is so strong that the birds establish a common singing area with marked boundaries in the middle of the colony. The largest songfest ordinarily takes place in the late afternoon.

Starting With Zebra Finches

In captivity, Zebra Finches don't have stringent requirements, and they will certainly start breeding given normal care and not subjected to unusual, stressful conditions. They are easy to please. If you don't provide nesting boxes, for example, they find a suitable nesting place on their own. For the beginning breeder, gray Zebra Finches with natural coloration are recommended. They have long been known for their strong constitution—and they are inexpensive. The various color variations and mutations are more likely to run into problems if there are difficulties, and they are

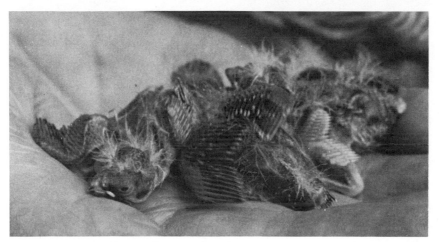

A handful of baby Zebra Finches.   *Donnelly*

A family grouping of Zebra Finches showing the hen at the left, the strikingly-marked cock in the center and one of their offspring at right.   *Sloots*

relatively expensive. An exception to this are the white and fawn varieties. Still, it's best to commence initially with the basic gray types.

Select birds thoughtfully. Take a pair of birds in complete good health with good feathering and without deviations in beak or legs. You can tell sick birds by the way they put their heads in their feathers to sleep all the time. They have hurried breathing and teary eyes. Look for a beak that is solid red, keg shaped, and solid. Both sexes have a red beak, but the female's is slightly paler.

Be sure the birds are old enough. You are dealing with a juvenile if the beak is black or has a black background. Remember, birds *must* have the full adult coloration before they can be bred.

Also, don't get really old birds, which are not good breeders. You can recognize these senior citizens by the color of their beaks. It will be red, but with a gray-white background. They look as if their beaks are coated with plaster.

Don't cut corners. If you want to breed good, even excellent birds, watch for the following characteristics in your breeding pair:

1. Uniform, black-white markings on the tail.
2. A solid, virtually straight, and—most important—not too slender back.
3. Well-pulled-up, well-connected wings.

The male, especially, should have a good outline on the breast. He should also have brilliantly colored flank stripes, with equal round, white spots placed near each other.

Don't start off buying a large number of breeding pairs. Begin with a pair or two—three at the most. Then you can devote proper attention to each. You will be better able to concentrate on the basic principles of care.

If you breed two pairs, don't put them together. Two pairs together invariably make trouble. One pair (or three) is easier.

In selecting mates, keep good records and individual identification to avoid inbreeding. You need fresh blood. Particularly avoid brother-sister matings, or matings with parents, grandparents, nieces, nephews, or cousins. Good breeders avoid pairing related birds if at all possible to avoid genetic problems.

Another piece of advice, which really applies to all bird purchases: Never buy birds during a severe cold spell. In carrying them home, the birds can get sick, no matter how well you protect them. If you do transport birds at low outside temperatures, don't place them in a warm room immediately after you arrive home. Air in the lungs, in the air sacs, and bones would expand too rapidly, putting an enormous stress on the birds' bodies. Birds so exposed almost always die, and most painfully. So, let birds warm up gradually. Even with precautions, however, I strongly advise against moving birds in winter.

102

If you transport birds by train or automobile, wrap the travel cage completely, except for a small section on the front to allow air and light to enter. Don't try to keep water in the water dish; it would just spill out. Instead, furnish some soaked bread or a wet sponge to slake their thirst en route. Some seeds in the seed dish are okay.

Even under normal temperatures, birds that arrive home from a trip still need some special care. After all, the constant shaking and rattling of the cage tend to make the birds withdrawn, even frightened. So, leave them undisturbed in their travel cage for at least three hours after arrival to allow them to recover from the trip. Check that they have seeds in their seed dish and put water into their water dish. Don't give them cold water straight from the tap—let it reach room temperature. Cold water could cause intestinal troubles and other ills because birds tend to get relatively hot in their wrapped travel cages.

Housing

Use a good cage or aviary. There is no objection to keeping several species of birds in the same aviary with Zebra Finches. When the other birds start their mating dance and other mating behavior, they stimulate the Zebra Finches to follow suit. Take care, however, that the inhabitants are compatible. Birds won't breed if they are disturbed by belligerent individuals who upset nests and fight about room on perches for sitting or sleeping.

Once you're more experienced in breeding Zebra Finches, you can successfully keep a larger population. You can keep them in an aviary—provided it is roomy—or in a so-called colony breeding cage with a capacity for three to six pairs. Here, again, you could mix species if you like.

Breeding birds in a colony cage is generally easier and more spontaneous than using a breeding cage that holds only one pair. The colony cage allows the birds some freedom in selecting their mates while a mating selected by the breeder alone can cause problems. If the breeder puts together a pair that don't get along together, brooding and hatchling care can go radically wrong.

Colony breeding cages most closely approach natural breeding conditions. I've mentioned that Zebra Finches tend to mate for life in Nature, and this trait is maintained quite strongly in colony cages or aviaries. Colony breeding also follows nature in that Zebra Finches breed in groups, where the actions of one pair stimulate similar behavior in other pairs—such as pairing and nest building.

A colony breeding cage also has its disadvantages. It should be used only when one is absolutely sure there'll be no problem about color inheritance. In a colony cage, a newly acquired mutant bird can easily mate with the wrong bird, so that the expected new mutation doesn't materialize or turns out disappointing. Furthermore, if a new mutation appears spontaneously—and this can happen at any time—there is another problem. To preserve the mutation it is wise to back-cross the mutant bird to one of its parents. This is quite difficult to do in a colony arrangement where the breeder can not know which bird was the mother or father of the mutant.

That doesn't mean, however, that the breeder can exercise no control at all over the genetic composition of the hatch raised in a colony. On the contrary, he can be quite successful in stabilizing new mutations. He can expect excellent results in maintaining certain purely heritable color variations, such as fawn, in Zebra Finches.

The fact that Zebra Finches in Nature mate for life has its advantages, but it does set up problems if you want to remove one bird of an established couple, perhaps to replace it with a bird of a different color. The new mating may succeed, but not completely. One must remember that if the members of an original pair are very attached to each other, a strange bird breaking up the relationship is traumatic. If for some reason you must separate an established pair, make certain that the bird you remove isn't placed anywhere near its former mate. Not removing the bird far enough is a frequent error and it leads invariably to ruining the future of the newly-formed couple. It is best to wait until the end of the breeding season to break up a couple. Making changes during the season causes a disturbance even among birds who have been rearing young satisfactorily.

Things are different if you use breeding cages that house just one pair. Then you can try substituting a bird during the course of the season. The birds need some time to get used to one another. Then it is possible that the newly matched pair will still raise one or more sets of young the same season.

If you separated a couple that weren't raising young, you can try to match the partner that you removed with a new mate as well. In many cases, the new mating will be successful. If not, you probably have an unsuitable bird on your hands, one not worth working with any further.

Unusual Traits

Not infrequently, a male Zebra Finch suddenly and rapidly starts building a sleep nest for his young. This is not an abnormality, but a spontaneous reaction to a necessity that is very difficult to define. It may be triggered by the fact that the keeper—or Nature in certain cases—overlooked or neglected something. The bird responds by some self-preserving instinct related to the entire breeding process, which seems to us strange and inexplicable.

Another unusual development is layered nests, which occur at times,

Black-breasted Gray Zebra Finch. *Photo by Author*

White male Zebra Finch. *Photo by Author*

Crested Gray Zebra Finch. *Photo by Author*

especially in large aviaries with a good-sized population. There are several possible causes.

First, you may have a female with poor blood circulation, which prevents her from laying a normal clutch. She stops with two or three eggs, which stimulates the male to build a new nest on top of the old one. Often the female helps him with amazing intensity. The result, of course, is that the first eggs are wasted. Poor blood circulation can cause the problem to repeat itself a second and third time, and in each case a new layer is added to the nest. The best way to avoid this problem is to select healthy birds for breeding. Birds that are not too young—about 14 months—should give good results.

Secondly, layered nests can result from a surplus of nesting material in the aviary. In that case, nest building is obviously overstimulated to such an extent that the birds cover up their first nest—often already with a complete clutch inside—and build a second nest on top. It is almost always only the male who gets caught up in this activity. The female gets involved much less. In the wild, this situation probably occurs only rarely. "Natural" nests are constructed loosely with a minimum of building materials.

A third cause can be a poorly matched, unnaturally formed pair— typically one formed by a breeder who separated a bird previously coupled with another mate. Especially when a recently separated female still can see or hear her original mate, layered nest building can result.

A fourth reason can be nestboxes too big for Zebra Finches. Birds tend to want to fill up nest boxes to the very entrance. If you must use boxes too deep for your finches, fill them up yourself with nesting material up to 1½ inches (4 cm.) under the entrance hole.

A further case of layered nests occurred in the wild. In Queensland, Australia, I visited a small colony of 23 nests after the sudden onset of a serious drought, and I found all 23 nests covered with new nesting material. When I returned to the colony some time later, 19 nests had been abandoned, while the other four were filled with new eggs. The rest of the birds obviously had built new nests elsewhere.

Some final observations about sleeping nests, which don't really differ much from brooding nests. They tend to be sloppier in construction and don't differ much between aviaries and in the wild in Australia.

The male has an unusual way of leading his young to the sleeping nest. He puts something in his mouth with which to entice the hatchlings. In Australia, I saw males with termites in their mouth direct young birds to different sleeping nests—of which there may be several. As far as I could tell, they don't worry at all whether the birds they lead to shelter are from their own brood or not. I saw one male lead 12 young to different sleeping nests when I knew he had fathered only four!

While sleeping nests always are used by wild birds, aviary birds may skip building them during a nice summer with warm days. Instead, they direct the young to a protected area or assigned tree branch. It helps to be aware of the sleeping place a male has selected in the aviary to be sure that no cats can reach it. That's one reason to recommend double netting around the aviary.

The Bengalese occurs in various colors, and there are even some crested varieties.

Photo by Author

7

Breeding Bengalese

THE BENGALESE of the aviary have no real counterpart in the wild. The Japanese have crossed and recrossed their antecedents in ways that are hard to trace. Breeders, ornithologists, and especially Japanese aviculturists have spent years trying to backcross Bengalese to determine their original ancestors. The hypothesis is that the birds are the chance result of a mating of Sharp-tailed Mannikins and Striated Mannikins (*Lonchura striata acuticauda* and *L. s. striata*). Allen Silver believes that, in addition, the African Silverbill (*Euodice malabarica cantans*) was involved in the development of the Bengalese.

Bengalese are good breeders if given enough room. They usually are quite passive and don't stand up to other birds in the aviary. They will allow nesting material to be stolen out of their nest box, even if they are in the midst of building or brooding. They let themselves be pushed around to the point of giving up their whole nest to other birds. There are individual Bengalese which will stand up for their rights, but they are the exception.

In short, Bengalese need support from the hobbyist to breed successfully—not because the birds are unwilling, but because they don't get the chance. So, provide these birds a breeding cage of adequate size or an aviary with plenty of space.

Many aviculturists consistently get 15 to 20 young from a pair of Bengalese per year. I personally think this is overdoing things, and the birds should not be taken advantage of. Unless you intervene, they will continue breeding winter and summer. That's not advisable because the female would go under from egg binding and general exhaustion. Select a breeding time for them, either in spring or summer, that fits your schedule. The winter months also can be used if you can supply a spacious, well-lit indoor aviary room or breeding cage.

It's hard to sex Bengalese. Only the song of the male—a rather attractive rasp—can help sex the birds successfully.

To further confound a would-be breeder, two females put together act like a breeding pair. They even build a nest, lay eggs—infertile ones, naturally—and take turns brooding. Even two males together can act like a pair. They don't sing, build nests with brooding chambers, and will sit for days on the empty nest as if they were brooding.

So, when you buy a pair of Bengalese, be sure your sales contract permits you to exchange a bird if they don't turn out to be an effective couple. To spell out the exact guarantee is good business for both buyer and seller.

If you pick a reliable dealer and have a little luck, you will, however, be able to acquire a real breeding pair. The normal brooding period for Bengalese is 20 to 21 days. After another 20 days, the young are ready to leave the nest, but the parents still continue to feed them for a long while.

Bengalese make exceptional foster parents for more costly aviary birds. Eggs from many species (especially Australian grass finches) can be given for adoption to Bengalese, which will reduce the risk of mishaps. It doesn't matter if the eggs have been brooded by the original parents for a while. The Bengalese will not be upset if the eggs they are brooding hatch after they have been in their care only, say, five or six days.

Even young hatchlings abandoned by their true parents can be entrusted to Bengalese for further care if the Bengalese have been raising young of more or less the same age and stage of development.

Foster nestlings of Bengalese should be returned to their natural parents once they become adults. Or else, these young birds will always want to stay near Bengalese and won't associate with their own kind. Prevent such imprinting if you plan to breed the birds later on.

Feed Bengalese the way you would Zebra Finches and other small tropical birds. Bengalese should also be given cod liver oil to ward off egg binding and other ills. Mix three drops into every two pounds of seed or quart of water. They also need cuttlebone and lime all year long.

Be sure to furnish clean water for drinking and bathing. Replace water at least once per day.

Note that Bengalese are like nuns or mannikins (*Lonchura*) in that their nails can grow exceptionally fast. Keep an eye on this to avoid accident or injury. If you furnish plantings of reed or put rough stones in the aviary, the birds generally will take care of the problem themselves.

8

Breeding Fruit- and Insect-Eating Birds

FRUIT- AND INSECT-EATERS generally are bred in heated indoor aviaries or in "greenhouses" behind glass. In these quarters, the breeding season can start in winter as well as in spring or summer. You can help the birds select one of these seasons, and if the result is an out-of-season hatch (relative to the natural bird population in your area) that shouldn't be a problem.

The challenge of winter hatching is that the birds need a rich assortment of insects and fruits, especially while brooding and raising their young. Since insects are difficult to come by in winter, you'll probably have to raise your own!

True fruit eaters need raisins, currants, oranges, bananas and similar material, especially at breeding time, regardless the season of the year. Special reading on these birds is a good idea and a list of preferred reading material will be found in the bibliography.

It is possible, but more difficult, to breed these birds in outside aviaries. These have to be roomy, with a variety of plantings, and equipped with an absolutely frost-free sleeping enclosure.

When you decide to breed these birds, remember that many species originate in the Southern hemisphere, where the seasons are turned around from our point of view. By nature, these birds expect to breed during their summer season, from mid-October to April. So if you plan to breed during the northern summer, the birds have to make a six-month change—an adjustment that requires time and special care. Generally I would say the process takes several years. The message here is: Buy only completely acclimated birds. They need to be accustomed to your local climate and—note this well—to locally available feed if breeding is to be successful.

Blue-eared, Red-flanked or Pleasing Lorikeet (*Charmosyna placentis*); approx. 7 inches (17 cm). Habitat: New Guinea, the Aru Islands, and the Western Papuan Islands.
Photo by Author

Purple-naped Lory (*Lorius domicellus*); approx. 11 inches (28 cm). Habitat: Ceram and Amboina; introduced to Buru.
Photo by Author

9

Breeding Parrots
and Parakeets

$\mathbf{B}$REEDING LARGE PARROTS, like cockatoos or African Greys, in captivity had long been considered impossible. The prevailing opinion and evidence held that breeders could count on success only with Budgerigars, Australian parakeets of the genus *Platycercus,* and—more recently—love birds (*Agapornidae*) and South American species. That this limitation can be overcome has been definitively proven during the last few years.

One of the primary keys to success in breeding parrots in captivity is to remember that a parrot does not tolerate the company of any bird except its mate during the breeding season. So, as a breeder, you must be sure to provide a spacious aviary—as large as possible—exclusively for the breeding pair. They'll need the space, if only to be able to get away from one another at times. The relationship between the breeding pair itself is definitely not all sweetness and light.

THE PARROT FAMILY:
NESTING AND BREEDING HABITS

Parrots won't breed unless their accommodations and care are optimal. Install nesting boxes for them in the run, not in the sleeping area or in the covered portion of the outside aviary. Their food must be of prime quality.

The breeding pair has to be in excellent health and the optimum age. Most parrots mature sexually rather late—at 2½ to 3 years of age, and sometimes considerably later. It makes no practical sense to put a couple of one- or two-year-old birds together and expect them to breed!

Once parrots start breeding, don't be surprised if they do it in the strangest places. I know of a case where a pair of African Greys selected an old oaken bookcase full of old books somewhere in an attic to start nesting and breeding. The birds were an unlikely couple—the female was four years old and the male was 12. They had been getting some exercise daily in the attic and one day they decided on their own to add a little spice to their life.

Their first hatch produced two young. The next year, they raised a singleton. They kept on breeding several more years, with a hatch of from one to four young. That's just one example to cite among many.

Not all parrots require nesting material. Love birds and Monk parakeets do; other parrots and parakeets don't. Nevertheless, a handful of wood shavings or peat moss put into their nesting place helps stimulate breeding.

Hookbills like to nest in a hollow in a rotten log or in nest boxes. The nest should not be too large—just big enough to accommodate mother and one to six young. The entrance hole should be so small that the adults have to squirm their way in with some difficulty. If the hole isn't big enough as furnished, the birds themselves will enlarge it with their sharp bills.

The parrot family's larger members take a month or longer to brood their eggs (usually two or more). The smaller species take about three weeks.

Now, a few specific remarks on the various species. The first birds I will discuss are the Australasian avifauna.

Lories: There are 11 genera of the Lory family (*Loriidae*). They include 55 species and 88 subspecies.

In Nature, lories leave their colonies at breeding time and separate as pairs. They nest in hollow trunks or branches, generally covering the bottom with rotted wood. Eggs (usually two) are laid a day apart.

The large species brood 23 to 28 days; the smaller ones, about 21 days. The female is the sole brooder, although the male often joins her on the nest at night. And in some species, he takes part in actual brooding, as in the *Vini* genus and in the Collared Lory (*Phigys solitarius*).

Most hatchings occur during the Northern winter. It's essential that captive birds be kept in a roomy, inside aviary that is heated moderately. Lories always should be housed in pairs. Don't mix them with other birds, whether they are related hookbills or not.

If you want tame birds you can handle, you have to start raising them by hand as youngsters.

Cockatoos: The subfamily of cockatoos has five genera, containing a total of 17 species and 28 subspecies—among them, the universally-famous Cockatiel. They are, as a group, an interesting bunch of hookbills.

The diet of cockatoos includes root vegetables, nuts, seeds, berries, leaf buds, and young leaves. (Cockatiels, for example, just love privet leaves.) In addition, cockatoos eat insects and larvae, especially at breeding time. They catch insects in rotting wood and in soil as they dig for roots.

In Nature, cockatoos prefer to nest high in trees, often eucalyptus, which grow close to water. They use hollow trunks and branches, which they decorate with bark from the same tree. Bigger species lay one egg; even when a female hatches two young, she raises only one and lets the other die. Small species lay two to five eggs.

Eggs are brooded in turn by both sexes. There is a definite division of labor. The male broods by day and the female at night. While she broods, he stands guard nearby. Eggs hatch after 35 days for big species and about 30 days for smaller ones.

Parrots: The parrot family (*Psittacidae*) has four subfamilies.

The first subfamily is the *Nestorinae,* which includes the famous Kea and Kaka of New Zealand *(Nestor notabilis* and *N. meridionalis).*

The Kea and Kaka eat about the same diet as cockatoos, but, unfortunately, they have gained a bad reputation for eating carrion. Both species nest on the ground, or between branches and moss, or under overhanging rock formations.

Eggs (four to six per clutch) are brooded exclusively by the female, although the male constantly hovers nearby. Brooding takes a month. After hatch, the male brings in food. He presents this to his mate, who in turn feeds the hatchlings. After they are a month old, the male feeds them directly. After 2½ months, the young leave the nest.

The second subfamily of parrots contains the pigmy parrots (*Micropsittinae*), which have six species and 20 subspecies. Pigmies really are small, about four inches (10 cm). They spend a lot of time on tree trunks and eat mosses, fungi, and possibly an insect now and then. Pigmies also may eat berries, other small fruit, figs, and dates. In the wild, they nest in termite nests located in trees. They use these nests for brooding and sleeping. The termites close off that part of the nest occupied by the birds, and neither birds nor insects seem to cause the other harm.

The true parrots (*Psittacinae*) constitute the third subfamily, with a record 45 genera, 232 species, and 456 subspecies.

Finally, the Kakapo (*Strigopinae*) subfamily from New Zealand, forms the fourth subfamily. In the wild, these birds have just about died out. A count in 1978 found only eight individuals remaining because man "cultivated" and thus destroyed their habitat.

Now for a look at the Afro-asian families.

The Vasa Parrots (*Coracopsis*), which encompass two species and

The Senegal Parrot (*Poicephalus senegalus*) is the most popular member of this family of pleasant birds. Given acceptable conditions, the species breeds readily in captivity.
Photo by Author

The Galah (*Eolophus roseicap-pilus*) makes a very delightful pet and will also breed easily in captivity given the right conditions.
Photo by Author

seven subspecies, occur on Madagascar (Malagasy) and nearby islands. Quite well-known is the genus *Psittacus,* to which the African Grey parrot belongs. It has three subspecies, including the Timneh. The Greys live in the richly-forested lowlands of Central Africa and in the mangrove forests of East Africa.

They, too, nest high in trees in holes with rotted wood as a floor cover. They lay two to four eggs, which are brooded 30 days. At first, the female feeds her hatchlings singlehandedly.

Parrots of the genus *Poicephalus* fall into nine species, including the well-known Senegal Parrot (*P. senegalus*). Members of this genus also nest in tree hollows. The clutch of two to four eggs is brooded 25 days, and the young are fed in the nest 2½ to 3 months.

The genus of love birds (*Agapornis*), with nine species and nine subspecies, is quite well-known. Most of the species nest in tree hollows, but *A. pullaria* prefers termite nests in a tree. *A. roseicollis, A. fischeri,* and *A. personata* are colony breeders and look for nesting places principally in rock crevices or in nests previously used (and then abandoned) by weaver birds. They do, however, also nest in tree hollows.

A. cana, A. taranta, and *A. pullaria* transport strips of bark, grass, and other nest material between their flight feathers, while *A. roseicollis* also carries nest material in its tail feathers. The other species transport nest material in the beak.

All species lay three to six eggs, and the female usually starts brooding after laying the first or second egg. Brooding takes an average of about 24 days. The female broods alone, although the male does keep her company on the nest, often in the entryway. However, males of the *A. fischeri* and *A. taranta* species also take part in brooding. After five to six weeks, the young birds leave the nest, but for the next two weeks or so, the young return to the nest for the night.

The "hanging" parrots of the genus *Loriculus* get their name because they roost head down, like bats. There are 10 species and 22 subspecies.

They prefer to nest in tree hollows, usually quite deep ones. The birds close the entrances with strips of bark, and blades of grass, or other vegetation. In these species, the female once again carries building material for the nest between her feathers.

Eggs are laid in clutches of two to four and hatch out after 19 days. The male doesn't contribute to the brooding of the eggs, but both parents feed the young after they hatch.

The last genus of the Afro-asian parrots, *Psittacula,* contains 13 species and 26 subspecies. Many of them are super attractive in appearance, such as the Emerald-collared parakeet (*P. calthorpae*).

They use tree hollows for nests, including old woodpecker holes, which they enlarge with their strong beaks. They also nest in cracks in walls, under roof tiles and similar places. They usually maintain their colony association, which means that several pairs often nest in the same tree.

Females lay three to four eggs, which usually are brooded by both sexes, although sometimes the male does not participate. The smaller representatives brood about 21 days, the larger ones about 27 days. After five to six weeks, the hatchlings leave the nest.

Among the South American avifauna, the most spectacular is the beautiful macaw genus, *Anodorhynchus* (Hyacintara), which contains three species. Then there is the Spix's Macaw genus (*Cyanopsitta*), with just one species; and the genus *Ara* with 13 species and 10 subspecies.

The majority of the species brood in the hollows of dead palm trees. They look for sites high in the trees, so that it is difficult to study the inside of the nest closely.

The female lays two to five eggs, which she broods for about 25 days. The male comes to keep her company on the nest during the evening and night hours. During the first month or more, the female also feeds the young without help, but the male starts assisting her after that. The hatchlings leave the nest at 12 to 14 weeks of age.

The genus *Aratinga*—the conures—is increasingly getting attention from fanciers. We know of 18 species and 45 subspecies, which breed in a variety of places. They dig holes or use rock crevices, tree hollows, termite nests and tunnels in river banks for nesting sites.

The female lays three to eight eggs, which she alone broods for about 27 days. In this genus, also, the male will often join his mate on the nest for the evening and night. The young leave the nest at about 50 days of age, but after that, the parents still care for their young a considerable time.

The genus *Nandayus* has the Nanday Conure as its sole representative. Another single-species genus is *Leptopsittica,* where we find the Golden-plumed conure *(L. branickii),* a native of the Andes Mountains. The Yellow-eared Conure *(O. ecterotis),* sole member of the genus *Ognorhynchus,* also lives in the Andes.

The Thick-billed parrot of the *Rhynchopsitta* genus lives in Mexico. The bird broods in hollows high up in trees, and it often makes use of woodpecker holes.

The Patagonian Conure (*Cyanoliseus patagonius*) and its three subspecies belong to the genus *Cyanoliseus.* They usually nest in holes they dig themselves, with a passage that sometimes extends more than 10 feet (3 meters). The female usually lays two eggs.

Then there is the genus *Pyrrhura,* containing 18 species of conures with 37 subspecies, many of which are kept in captivity today. In the wild, they live on fruits, berries, blossom leaves, insects (and their eggs and larvae), and a rich assortment of seeds. The nest is usually in a tree hollow. The clutch consists of two to four eggs on the average, but as many as eight have been recorded. The female alone broods the eggs, which hatch in 22 to

The Yellow-winged Blue-fronted Amazon is closely related to the more familiar Blue-fronted Amazon and is typical of the genus *Amazona*.

Photo by Author

Illiger's Macaw (*Ara Maracana*) is a representative of the "dwarf" macaws. He is well-established in aviculture and there has been some captive breeding.

Photo by Author

28 days. After about two months, the young leave the nest, but they are still fed thereafter, principally by their father.

The genus *Enicognathus* contains two currently quite popular species, the Slender-billed Conure and the Austral Conure. Both come from southern South America.

The genus *Myiopsitta* contains the well-known nest-building Monk Parakeet, of which four subspecies are known. Its copious nest is made out of branches, twigs, leaves, grass, and a variety of other materials. Thorn-bearing branches are used to help keep out enemies. The birds are colony breeders.

Another South American genus, *Bolborhynchus,* contains the well-known aviary species, the Lineolated Parakeet (*B. lineola*) and the Aymara Parakeet (*B. aymara*). There are a total of five species, all of which nest in river banks. The female lays five to seven eggs.

The *Forpus* genus consists of small, colorful birds known as parrot-lets. It counts seven species and 19 subspecies. They live on fruits, berries, leaf buds, blossoms, moss, and grass seed. They nest in tree hollows and similar places. The eggs, laid in clutches of two to seven, are brooded for 18 days. The hatchlings leave the nest after four to five weeks, but they continue to be fed by the parents after that.

The *Brotogeris* genus of slender-billed parrots has seven species with 15 subspecies. The two sexes are hard to differentiate in looks. They have a very typical narrow, pointed and protruding bill, with a wide, rounded notch in the upper mandible. Their diet is omnivorous. They nest in trees—sometimes in termite nests—and almost always in rather large groups. The female lays four to six eggs in a hollow, and she broods them for about 26 days. After about two months, the young leave the nest.

The genus *Nannopsittaca* is native to Venezuela and British Guyana. It consists of a single species, the Tepui parrotlet.

The *Touit* genus has seven species and six subspecies. They live principally on vegetation—buds, blossoms, flowers, nectar and berries.

The genus *Pionites* encompasses the Black-headed and White-bellied Caiques. The female lays two to four eggs, which she alone broods for about 28 days. The hatchlings are about two months old when they leave the nest.

The following genera have only a single species: the Vulturine Parrot (*Gypopsitta*); the Short-tailed Parrot (*Graydidascalus*); the Hawk-headed Parrot (*Deroptyus*); and the Purple-bellied Parrot (*Triclaria*).

The genus *Pionopsitta* has six species and four subspecies. Among these typical forest dwellers, the female alone broods the eggs.

The genus *Hapalopsitta* has two species and five subspecies.

The well-known genus *Pionus* has eight species and 15 subspecies, which are spread over a wide range of habitats. Their principal food consists of berries, seeds, blossoms, and all kinds of fruits. They have a

familiar, comical profile of Toco Toucan, ♂, (*Ramphastos toco*). The enormous beak is really ...remely light and is an important survival modification.

Photo by author

Above: Diamond Finch, ♂, *(Stegan-opleura guttata). Right:* Parson Finch *(Poephila cincta).* Photos by author

Orange Bishop, ♂, (Euplectes orix franciscana)

Photo by author

Red Bishop, ♂, (Euplectes hordeaceus)

Photo by author

Above: Golden-breasted Waxbill, *(Estrilda subflava). Left:* Orange-cheeked Waxbills *(Estrilda melpoda).*

Photos by author

Fischer's Lovebird *(Agapornis fischeri)*.   *Photo by author*

Gouldian Finch, ♂, *(Chloebia gouldiae).* This is one of the most colorful and popular of all finches kept in aviculture. *Photo by author*

Bronze-winged Mannikin *(Lonchura cucullata).* *Photo by author*

Chestnut-faced Starling
(Sturnus malabaricus).
Photo by author

Little Masked Weaver, ♂, *(Ploceus luteolus).*   Photo by author

Lazuli Bunting, ♂, (Passerina amoena).   Photo by author

Varied Bunting, ♂, (Passerina versicolor).   Photo by author

reputation for making damaging forays on corn and rice fields. The female lays three to four eggs, which are brooded for about 25 days. Hatchlings are fed by both parents and leave the nest after about two months.

The last of the South American parrots are the Amazons—the genus *Amazona*—with 27 species and 42 subspecies. Most Amazons are green. They spend most of their time in trees, and move about in small flocks. They can cause considerable damage to farmland and gardens, so that they are unfortunately being killed off. Many are protected by law, a strange development because the destruction of their habitat continues unabated.

Amazons gather in large flocks during the evening to return to their established sleeping areas, from which they can range a surprising distance. Their foraging expeditions yield all kinds of seeds, fruits, berries, nuts, leaf buds, blossoms, and young twigs.

The nests of Amazons are made in a natural tree hollow. The female lays two to four eggs, which she alone broods for about 27 days. The male, however, regularly comes to look and constantly keeps watch somewhere in a dense tree crown near the nest. It is almost certain that he feeds the female on the nest.

Captive Breeding

Breeding fancy parakeets and parrots has expanded strongly in recent years. It's a well-known fact that the currently intensive captive breeding enterprise has saved a number of species from extinction. Captive breeding is of such importance that some species have more individuals living in captivity than in the wild.

We could conclude from this success that raising fancy parakeets is easy. However, as a breeder you should realize that the success rate is not equal for all species. Still, if you follow the directions which follow, most of the birds you keep will regularly furnish you with young ones.

The Australian "fancy" parrots are probably the hardest to breed, while they also are perhaps the most popular birds of all. And yet, if you have a good breeding pair, you can get quite good results. As a breeder, you should not be afraid to try an experiment or two.

Don't start breeding smaller psittacines before they are at least eight months old. Larger species should not be bred until they are 10 to 12 months old. Most species should not be bred after they are four to five years old, although experience hs shown over the years that this general rule has numerous exceptions. I know of birds that have raised good broods at 10 years of age and definitely didn't have to take a back seat as breeders to a younger generation.

Avoid inbreeding as a matter of course, except if you want to preserve certain mutations through this method—or if you want to bring them about. This was done by Dr. L. A. Swaenepoel of Lembeck, Belgium, with Ringneck Parakeets (*Psittacula krameri*).

Red-vented or Blue-headed
Parrot (*Pionus menstruus*);
approx. 11 inches (28 cm).
Habitat: the greater part of
tropical South America.
Photo by Author

Peach-faced Love Bird (*Agapornis roseicollis*); 6 inches (15 cm). The largest and the most
popular of all love birds! The yellow mutation is widely-known. *Photo by Author*

Keep a careful eye on your birds at all times. That's important in order to know exactly what the birds need from you. Especially monitor their food supply and breeding boxes.

Housing

Also of key importance is housing breeding pairs in separate quarters. Again, you can expect the best results if you put each pair in a separate aviary. (There are a few exceptions, which will be mentioned further on.) Housing two or more pairs of the same or related species in one aviary is very unwise; it only leads to fights between the males, infertile eggs, and other troubles. Even in a really roomy aviary with large compartments, you still should avoid putting two couples of the same or a closely-related species next to each other.

"Closely-related species" can include a whole range of birds, not only birds in the same genus. Any birds that *could* interbreed, and under the artificial conditions of captivity, many species have done so, qualify as "closely-related." For example, it is possible to cross the well-known Crimson-winged Parakeet or Red-winged Parakeet *(Aprosmictus erythropterus)* with the Rock Pebbler, or Mountain Parakeet *(Polytelis anthopeplus)*; as their scientific names indicate, they belong to different genera. It obviously pays to be extra careful.

Birds that are only "closely-related" don't need to be kept as far apart as truly related species. Even though most breeders have to fight space limitations, these remarks should not unduly alarm. Practically speaking, this advice may be difficult to follow and shouldn't be taken as an invariable absolute. A good rule of thumb is never place birds of the same species next to or with each other in an aviary; place birds of the same genus (or species) as far apart as possible. Within this basic rule, do what is most possible and practicable.

Another reminder: Keep the area around the aviary as quiet as possible. I've always had a dog or two while I've been breeding birds, and it's amazing how the birds got used to them. But it's best to keep dogs away from the birds, cats must be considered a major hazard. (Are there *any* bird breeders that keep cats?)

Keep your own movements quiet, especially when you come close to tend to or observe your birds. They'll use the least disruption as an excuse to stop brooding. It helps to feed and water your birds at set times. They'll get used to your schedule quickly, especially if you sing or whistle softly when you have to be near or in the aviary. It is amazing how much this calms the birds.

Nesting in the Aviary

The various types of parrots and parakeets almost never use nest-building material. Only Monk Parakeets and love birds build a true nest—exceptions that surprise even professional ornithologists. So, breeders of fancy birds should supply them with brooding places, like nest boxes or nesting logs.

This requires special attention. Each breeding pair should have at least two different types of nest boxes available, so that they can make a choice. In other words, the breeder shouldn't force them to take or leave the one nesting place he provides.

Personally, my best results come if I give the couple four nesting logs of two different types. Two are hung in the outside flight cage and two in the night shelter. It's best to hang both types on the same wall. Then move both nesting logs to the opposite wall if the birds take no notice of the facilities as first provided. I give my birds about 10 days to decide if they like the nesting logs where I first put them.

You may ask why I don't provide a nesting log on *each* wall. The answer is that it's impossible to determine in advance which type of nesting place the birds prefer. Naturally, I make sure that I can check on the nest boxes on either wall without unduly disturbing the birds.

After several breeding seasons, the breeder knows the preference of his birds, of course. That simplifies the job in the future.

The best way to make nesting logs for the birds—especially the bigger species—is to hollow out a log. The Rosella species, for example, need a nesting log about 18 inches (45 cm) long with a diameter of about 10 inches (25 cm). A log that size is hard to carve with a sharp knife. Sawing the log lengthwise and then hollowing it out is easier and safer. The best way is to cut off a slab, then make the hollow underneath. Put hinges on the slab, and you have an easy way to look into the nest if you have to, or to clean it out.

The hinged slab may not be best if you place the log where drafts can blow through the cracks; it would be bad for the health of parents as well as young that use the nest. In those cases, glue the pieces of log back together. Get a good, tight bond to keep the log from drying out and splitting.

Nesting logs need an entrance hole about three to four inches (8 to 10 cm) from the top. The hole itself should be no more than three inches in diameter—and it must be *round*. Also, drill a few ventilation holes, perhaps 1/3 inch in diameter about 1½ to 2 inches (4 to 5 cm) below the top section of the log. That's important because a brooding bird may go to the nest opening for a breath of fresh air and block just about the entire hole with its body. The young, inside, thus have their air supply partially or wholly cut off. A nest holding more than two young, especially, can run out of fresh air quite rapidly.

You can buy nesting logs (generally made of birch) in stores, but for most of the fancy psittacines, these commercial nests are too small. You could try some of these commercial units for the Neophema Parakeets—the Turquoisine, the Splendid Grass, the Bourke's, and the Elegant Parakeets.

Love Birds (*Agapornis sp.*) are being developed in many beautiful color mutations. The birds perched on the nest box are Golden Cherry Peach-faced Love Birds. The bird at left is a lutino and the other is called an "Australian Cinnamon." Above them an "Australian Olive" perches on the wire.

Sloots

An aviary nest of a Fischer's Love Bird showing eggs and young. The design of this nest box allows for easy inspection and proper ventilation. *Sloots*

As an alternate nesting place, build the boxes out of boards. The size of the nesting place should be the same as suggested for hollowed-out logs. These boxes are easier to make and light in weight, so that there is less risk that they will come tumbling down after they are installed.

For the large, fancy parakeets—Barrabants (*Polytelis swainsonii*) and Australian Kings (*Alisterus scapularis*)—use large, standing tree trunks of at least five feet (1½ meters), but not taller than six feet (2 meters). Under the entrance hole three inches (8 cm) on the average, install several hooks and a strip of wire mesh to help birds climb out. Be sure, however, that you don't leave any sharp, loose or protruding wires that could injure the birds. Legs are especially vulnerable to this type of injury. Place the tree trunks at a slight angle to help the birds climb.

Even nesting places of this size can be hammered together from smooth boards. Be sure that the diameter from corner to corner is never less than eight inches (20 cm), nor more than 10 inches (25 cm).

Even though these birds use practically no nesting material, some like to fill the bowl of their nest with some grass, wool, strips of paper, willow bark, horse hair, feathers, jute fibers, moss, or rotted wood pulp. I like to furnish the last two items in a moist condition. If the birds don't use them on their own, I place them in their nest myself.

Incompatible Pair

You have prepared the aviary and the nesting places and have installed your newly purchased birds. What if, after all the preparations, the birds won't start breeding? First, recheck. You may not have furnished the right types of nesting places. Or the diet may be wrong.

If you still get no action, you may have bought the wrong birds.

First, you may have had a non-producing pair foisted off on you—one that travels from hand to hand, proving each time that raising offspring is not for them. Since this happens, be sure that you always buy birds with a guarantee, allowing you to exchange non-breeders.

Second, the vendor may have sold you a non-breeding pair in good faith—young birds that haven't had a chance to prove themselves. In that case, exchange one bird of the pair, preferably the male. A new mate sometimes brings the desired results immediately. Even hard-to-handle types that resisted breeding for years have settled down to breed when introduced to a new mate!

Incubators and Foster Parents

Some remarks are in order for those seeking an opinion on using incubators for fancy birds. Especially with costly species, I think a breeder is perfectly justified to use an electric incubator or other type of electrical brooding device. I don't consider this a crass intervention with Nature. The aviculturist would lose a handful of money if a brood would be lost. And, the financial aspect aside, artificial brooding can help prevent extinction of valuable species. I believe it of the utmost importance for aviculturists to restore what other people have destroyed. This is particularly true of certain valuable Australian parakeets that certainly would no longer exist if incubators were not used to help brood eggs.

One key reason for going to artificial incubation is the case where a female won't brood her eggs. Lack of brooding drive can be caused by a number of factors. Proper food and housing can improve this drive significantly, but even in aviaries where these necessities are amply provided for, some females make absolutely no overtures to brood their eggs.

Don't be hasty, though, about deciding you have a non-brooder on your hands. In most cases, a female starts brooding after having laid her third egg. But wait until she lays a fourth egg before taking action. When the clutch is complete and the female still shows no interest in brooding the eggs, it is time to take action promptly.

Remove the abandoned eggs carefully. Put them in a cool place. If necessary, you can keep eggs in storage safely for 10 to 12 days, probably even longer. If you're using a small incubator—perhaps a homemade one—there is no need to hold off on incubating eggs. I mention the possibility mainly for the breeder who wants to use a single, large incubator to hatch a large collection of eggs, perhaps from a variety of birds.

The use of an artificial incubator doesn't rule out brooding by a foster mother. However, I don't recommend foster mothers as *total* substitutes for artificial incubation. The expanded clutch may prove too much for the female to handle, and without incubators to fall back on, you could lose both clutches of eggs.

When using foster mothers for brooding, synchronize the artificial brooding of a clutch of eggs with the natural brooding of a female bird, preferably one of a species related to the bird whose eggs are in the incubator. After six days, inspect the eggs in the machine and in the nest. Destroy all unfertilized eggs, and replace eggs taken from the naturally brooded clutch with fertile eggs from the incubator. Be circumspect in this transfer, and wait till the brooding female leaves the nest for a moment for a bit of exercise or a bite to eat brought by her mate to a nearby spot.

Unrelated birds are not ruled out as foster parents, although related birds are the intelligent first choice. If no related foster mother is available, pick a bird of about the same size as the natural parent.

Red-rumped Parakeets (*Psephotus haematonotus*) make excellent foster parents. They have raised broods of young composed of eggs from four or more different species of birds in a number of cases. Other good foster parents for smaller species of birds are Bourke's Parakeets; they can also handle eggs of Red-rumped Parakeets and Paradise Parakeets

(*Psephotus pulcherrimus*). Rosellas of any species also take good care of eggs from any other Rosella species.

Even Cockatiels and Budgerigars can function as foster parents for the smaller species of fancy parakeets and love birds. However, I'd use Budgerigars only to help raise Turquoisines, and even then I'd rate the chances for success only about 50-50.

Love birds also make very good foster parents for Bourke's Parakeets, Elegant Parakeets, Turquoisine Parakeets, Blue-winged Parakeets, and related birds. Readers with a special interest in the breeding, care, and brooding of love birds are referred to the existing literature (see Bibliography).

As for the actual operation of incubators, I personally have had the best performance and the best end results from the small model electric incubator—the type where the egg tray lies horizontally and is made of fine wire mesh. A piece of rough-woven jute over the mesh will keep hatchlings from breaking their legs by sticking their feet through the wire mesh of the egg tray.

The brooder is heated by hot air that flows across the eggs from above and leaves the unit along one of the sides or through openings in the bottom. The temperature just above the eggs is kept round about 102 degrees F (39 degrees C). I get the best results by using a constant temperature of 103 degrees F (39.5 degrees C). A thermostat keeps the temperature constant.

If the incubator doesn't have a built-in thermometer, a separate thermometer should be acquired to check on the proper functioning of the unit. Take the measurement just above the eggs against one of the sides, so that the bulb of the thermometer is at the height of the eggs. The incubator itself has double walls for improved insulation and even distribution of heat. Use an ether capsule thermostat with this unit rather than a bimetal thermostat. The ether capsule thermostat is far more sensitive than the bimetal thermostat, which is popular because it is sturdy.

Birds turn their eggs several times a day while brooding, and if you use a small incubator, you will have to duplicate this job by hand twice a day. Don't worry about eggs cooling while you open the incubator doors. An airing like that is actually beneficial; it assures good development of the eggs. You can turn the eggs in the morning before you go to work and again 12 to 13 hours later, in the evening. Be careful not to crack the shells—use a plastic spoon. During your morning chores, also check whether there is enough water in the tray at the bottom. There always has to be water in the tray to assure adequate humidity. The incubator must be kept humid to keep eggs from drying out and to be sure that the young hatch properly. Many people check the humidity with a hygrometer, but this is not really necessary.

At the time the incubated eggs are ready to hatch, put them under a naturally brooding bird. You'll have to take away the eggs she is brooding, which often constitutes a loss. You can, however, put her eggs in the incubator, which, figuratively, completes a full circle. If you have more brooding birds, then you can distribute the eggs you removed among these other pairs. This presupposes that the other birds started brooding about the same time. It's nearly impossible to keep track of all this in your head, and to bring it off, you need to keep accurate written records.

It helps to have an incubator standing by in case a female suddenly abandons her eggs. Even if you notice this only after several hours have passed, you can still save the eggs. Eggs of parakeets are much more durable than is commonly thought.

If you have good, naturally brooding birds and no mishaps, you still need to help them a little. For example, the nesting log must be kept from drying out. Put peat moss in the bottom layer of the nest and cover it with another layer of rough wood shavings. The humidity is necessary so eggs are brooded properly and hatched easily. A successful hatch, in other words, depends critically on proper humidity in the nest.

Parakeets sometimes stir around in the nest linings that breeders furnish, or they work the lining out of the nesting place. Dr. H. D. Groen makes the same observation in his excellent book, *Australian Parakeets, Their Maintenance and Breeding in Europe,* Groningen, 1966. When the lining is removed, the eggs wind up lying on a bone-dry nest bowl. For some species, that's not too bad, but in most cases, the eggs would be ruined.

You can keep the birds from disturbing the nest linings by using a double-bottomed nesting place. Drill some holes in the upper "floor," and place a bowl of water between the two floors. Another method, which I don't prefer, is to lay some sod, moistened and well pressed, in the nest box.

Make regular observations of the birds in order to properly follow the brooding process. Brooding birds must be monitored but should be left undisturbed as much as possible, whether they are exotics or parakeets. Most tropical and subtropical birds really react poorly, whereas parakeets, with a few exceptions, take it better. If you are careful about it, you *can* check on them regularly without too much risk.

Disturbing birds, I believe, is justified only if absolutely necessary—to remove infertile eggs or dead young from the nest or to take away eggs if the female is brooding irregularly. Pick a time when circumstances are favorable. Don't chase brooding birds off their nest—sometimes they just won't come back. So wait until the female leaves the nest on her own, as when she takes some exercise on the wing or has the male feed her. Even then, the female may follow your motions and may refuse to return to the nest as a result.

Watch the bird's actions closely. If after some time, she doesn't return to brooding, put your trusty electric incubator to work. (Never leave eggs unbrooded for more than 24 hours.) In place of the natural eggs, put some stone or plaster eggs, which you ought to have on hand all the time. When

A clutch of Cockatiel eggs. *Photo by Author*

A seven-day-old Cockatiel. The bulging crop shows that this young one has just been fed. *Photo by Author*

you notice the female brooding these substitute eggs, you can bring her own eggs back from the incubator.

A female doesn't leave the nest simply because of an inspection by the breeder. However, this *can* happen, so you should be prepared.

It is fairly common for the female to stop feeding her young, and you should watch for such a possibility. If you don't act promptly, you could lose some valuable young birds. As indicated earlier, young birds can't do without food for long. They often eat more than their own body weight in a single day. The observation that a female isn't feeding her young generally occurs too late!

Crop Milk

Raising abandoned birds by hand is just about impossible if they are less than 10 days old. Even if you devote your full time to these recent hatchlings, you seldom succeed. You tend to them every hour, and at best they last 10 days before they succumb to intestinal upsets and similar problems.

A currently popular theory I've heard to explain this holds that the first few days of life, young birds are fed with so-called "crop milk." Some universities are in process of researching this theory at the present time.

Whether young birds absolutely need to receive crop milk is, however, subject to strong doubt. Dr. H. D. Groen (in the book cited before) states that young parakeets indeed need crop milk the first few days of life in order to grow up normally. He says it is obvious that the mother bird starts producing this crop milk at the end of the brooding period when she is expecting that young birds soon will hatch.

An experience Dr. Groen had with Bourke's Parakeets, however, shows clearly that this doesn't, in fact, happen. He had placed a young Red-rumped Parakeet in the nest of these Bourke's just after hatching out the little bird in an incubator. At that point, the Bourke's female had not yet completed laying her own clutch and had been brooding two days at the most. Nonetheless, she accepted the young Red-rump and the same evening already one could see some food in the crop of the youngster. During the next few days, Dr. Groen added two more Red-rumps. Not only were all three raised successfully, they also turned out later to be good, healthy adults. One can therefore conclude that they were not deprived in their youth. The only way the theory of crop milk could fit this case is to suppose that the mother bird produced crop milk at a point when she couldn't possibly have been expecting young.

I had a similar experience with a Cactus Conure (*Aratinga cactorum*), which I gave two young St. Thomas Conures (*A. nana pertinax*) to raise. In this case also, the Cactus Conure female fed the adoptive young long before her own eggs hatched. And when I was in Australia to study in 1965, I experienced another such case with an Elegant Parakeet female, who was given an amazing five Turquoisine Parakeet young to raise on her fourth day of brooding her own six eggs.

Exactly what the situation is with regard to crop milk is not clear at this point. Still, female birds must have something special to give newly hatched young, because they are very difficult to hand-raise.

Hand Feeding

If you want to hand-feed an abandoned youngster 10 days of age or older, feed it thin oatmeal. On the first day, feed only watered-down oatmeal. For the first five days, don't use milk. Afterwards, add some milk to the cereal. Cook the cereal and serve it lukewarm, about body temperature, otherwise the birds won't eat it.

I add a little sugar to the thin oatmeal if a little bird won't start eating. This can work wonders.

Offer food on a plastic or silver teaspoon or serve it with an ear syringe. The smallest birds can best be fed with a cordial spoon, which is smaller than a teaspoon, and thus handier. However, ear syringes, eye droppers, or commercially made feeding syringes should be used only the first few days. The thicker mixtures provided after that would clog these devices.

To feed a hatchling bird, put it in one of your hands in a way that enables you to hold its head gently between your thumb and index finger. Then, hold the spoon or syringe in your other hand. After each feeding, wipe the bird's beak with a soft flannel cloth dipped in lukewarm water. Under no circumstances should spilled bits of food be left on the beak or body of the bird.

The first few days, feed every three hours—that is, six times per day. Night feeding isn't necessary, because, after all, a mother bird also doesn't feed her little ones during the evening and night. So don't disturb your own night's rest needlessly.

I have found the best times for feeding are 7 and 10 AM and 1, 4, 7 and 10 PM. Some references suggest starting as late as 8 AM, but I find that an early morning feeding is more effective than a late evening feeding. Birds need rest, and research has shown that birds are hungrier at about 10 PM than around 11 PM, the hour at which the last feeding would fall if we started the schedule at 8 AM. In the end you must set your own hours to suit your schedule.

The first three feedings, as well as the last one, are oatmeal, sweetened with honey, cane sugar, or powdered sugar, if necessary. Feedings four and five, contain, in addition, finely crumbled rusk or other sweet biscuit.

Be sure that the crops of baby birds are well filled, but don't overstuff. Remember the theory that the little crops ought to be empty after three hours. If that isn't the case for the first few days, you may be feeding too heavily. Cut back right away to avoid digestive upsets and intestinal problems. For the first three days, go back to the basic warm oatmeal cooked in water only and feed it as warm as possible.

After the third day, start feeding some raw apple sauce, sweetened with powdered sugar or honey, in addition to the established menu. Warm up the apple sauce—the birds won't eat it too cold. On the afternoon of the fourth day, add finely ground raw carrots to the fare. Serve the carrots warm, but don't cook them, so as not to lose essential nutrients. Mix in finely cut lettuce and pieces of apple. At this point, add a few drops of cod liver oil—never more than two drops on a portion of two ounces. Even better than cod liver oil is calcigenol enriched with vitamins.

Starting on the fifth day, add egg shells ground into powder to the meal, along with a bit of cuttlebone. Be sparing with this. A half teaspoonful distributed over all feedings for a day should be more than enough.

Each morning, make up all the food you will use that day; you don't want to be cooking and mixing six separate times. Store the day's supply in a cool place to keep it from spoiling. Rewarm the food before each feeding, or the birds will refuse it.

Five feedings per day are enough after the 10th day. After 15 days, cut back to four feedings at 8 AM, noon, and 4 and 10 PM.

While you're handfeeding, house the little birds in a roomy, wooden box, 15½ inches square (40 by 40 cm). Place the box in a draft-free, waterproof, warm place. Your living room would probably serve fine.

At first, keep the feeding job for yourself. Later, you can let others share in it. The hungry little youngsters will remind everybody with their cries when they want to be fed.

After a few days, the foster babies are used to the feeding routine, and it will no longer take so much time to feed them. As soon as they see you approach with food and teaspoon, they'll open their little beaks wide. Feeding becomes, literally, child's play.

Don't think, however, that you're home free at that point. Soon another time arrives where you will have to be very careful. The birds become flighty. They don't want to stay in their box, they keep running away, and they refuse all the food you offer. My wife has called it quite appropriately their "adolescent rebellion."

When that happens, move the birds to a roomy cage with well-formed, manufactured roosts. These should be so placed that you can reach your teaspoon through the mesh or bars of the cage to feed the birds. That's important because, at that point, the birds won't let you catch and hold them. If you try to do so anyway, you'll get bitten, and that can hurt. This testiness disappears after a while because hand-raised parakeets tend to become exceptionally tame and affectionate.

Enrich the menu at that stage with a wide variety of *fresh* grass and weed seeds, rolled millet, and canary seed. Be sure that you feed grass and weed seed (plus grape millet, if possible) with the stem attached. The birds take the stem in their beaks and learn to eat seeds while toying around in this way.

Keep feeding the birds in their new cage for 14 to 20 days, several times per day. After that, you can consider your job as foster parent completed and you can release your birds into a roomy flight cage. If the cage is outdoors, pick a warm day to release the birds. Too drastic a temperature change exposes them to the possibility of pneumonia, or, in a less serious case, to a cold. After having taken so much time and trouble, you wouldn't want to lose your birds through a small, thoughtless act.

Take special care that the young birds spend the night in an inside shelter for the first few weeks. Your aviary needs a light inside shelter, situated higher than the flight cage because birds, especially parakeets, like to roost as high as possible. After all the birds are in the shelter, close the entrance door for the night until 8 AM the next morning. If the weather is bad during the first week, just keep the young birds in the shelter the whole time.

Hand-raised birds are tame and can be lots of fun to have around. In many species, however, tame birds often are not suited for breeding; I have found this to be particularly true for parrots. "Fancy" parakeets don't pose that much of a problem, especially if you don't try to teach them a large vocabulary or make them ride scooters or do other tricks.

Dr. Groen, a respected authority, said the following on this topic:

> I have in my collection a hand-raised, unusually tame male of the Golden Mantled Rosella species, who not only has bred well but also has helped feed and raise young. The experiences of Dr. Overlaender (Bad Honnef/Rhein) are important in that he reports that his hand-raised Splendid Rosellas and pennants consistently had offspring. This was the case even with one-year-old birds. His report is interesting because he found that the females became quite aggressive during the breeding period, especially when there were young in the nest. They bit him in the fingers every time he filled the seed bowls. After the breeding period, these birds again became as trusting as before.

I can add to these interesting facts with my own experiences and those related by my friends. I know of successfully-bred, hand-raised birds among the following species: White-eared Conure *(Pyrrhura leucotis),* Lineolated Parakeet *(Bolborhynchus lineola),* Tovi Parakeet *(Brotogerys jugularis jugularis),* Tui Parakeet *(B.x. sanctithomae),* Canary-winged parakeet *(B. versicolurus chiriri),* Nanday Conure *(Nandayus nenday)*— eight times, Cactus Conure *(Aratinga cactorum)*—twice, Jenday Conure *(A. solstitialis jandaya),* Ringneck Parakeet *(Psittacula kraemeri)*—eight times, Blossom-headed Parakeet *(P. cyanocephala rosa)*—three times,

A young Red-rump Parakeet *(Psephotus haemonotus).*   *Photo by Author*

plus many, many Australian *Platycercus* parakeets. Dr. Groen's book gives many more examples for those who are interested.

I don't furnish a detailed list of Australian parakeets because hand-raised young of these birds are well-known to be good breeders. During the three years I spent in Australia, I saw so many hand-raised birds breeding that I lost count.

Cockatiels also do well in this respect. I had repeated excellent results in aviaries I kept in the Netherlands, Australia, and the United States. (Dr. Groen has not commented on Cockatiels.)

Dr. Groen does have some reservations. He makes a very useful comment, I believe, when he says:

> Species of the *Neophema* genus are very hard to hand-raise. The estimate is that only 50 percent of hand-raised Bourke's, Turqoisines, Splendids, and Elegants reached adulthood. This estimate certainly is close to real experience in the Netherlands, and published claims of greater success by foreign breeders ought to be taken with a grain of salt. They don't go into details on their feeding methods.

The simple reason is that breeders outside the Netherlands didn't have then and don't now have any better results than Dutch breeders. When I checked with 82 Australian and 46 American breeders, I became convinced that they certainly didn't raise more than 50 percent to adulthood.

Baby birds abandoned by their mother are not unusual in Nature. The female suddenly stops feeding them even though she continues to feed her other young as before. Ornithologists still haven't discovered the cause for this sudden halt in feeding. Even in the wild, several Australian parakeets often stop feeding one or sometimes even more of their young. A similar behavior is also found among several smaller birds, like Cuba Finches. I imagine that this behavior represents a kind of Natural population control measure. It also could be that the parent birds instinctively sense there is something wrong with certain young ones . . . that these wouldn't grow up to be completely healthy anyway if they were raised to adulthood.

A sudden halt in the feeding of captive hatchlings bears watching because—in contrast to what may be the case in the wild—abandoned, captive birds almost always are healthy and normal. The parents apparently are deliberate about refusing to bring feed to certain young ones for some unknown reason.

Abandoned birds can be hand-fed. If you feed them well for several days, they will quickly catch up to their brothers and sisters in size. At that point, many times, the abandoned ones can be put back in the nest. The female ordinarily will resume feeding them, acting as if nothing had happened.

Start working with your birds at the right time in the spring. Don't give birds access to the various types of nesting logs before the end of March. At home, I stick to a precise date: March 25th. It may be a bit cold as yet to be expecting eggs, but remember that the males and females need some time to get used to each other. They also will make a thorough inspection of the aviary, especially if they were housed indoors during the winter. The song and dance of the courtship ceremony takes more time, so that egg production really doesn't begin until much later—not until the end of April. (At the earliest, laying could begin a week sooner.)

After reading detailed descriptions of how young birds are raised and bred, you may have gotten the impression that this is an easy job. Let me caution you against excess optimism. Don't forget that you are working with living matter. You need to give your full attention to the housing, feeding, and behavior of the birds. Eventually you will be able to sense almost instinctively if something isn't going right with the birds. If you don't (yet) have this gift, your daily attention is not only desirable, it is absolutely essential.

The pheasant family includes some of the most beautiful birds in the world. Keeping and breeding these birds makes a challenging, absorbing hobby.   *Sloots*

10

Breeding Ornamental Fowl

ORNAMENTAL FOWL, for the purpose of this section, include many members of the pheasant family, such as Ring-Necked pheasant, pea fowl, quail, partridges, bantam chickens, and turkeys.

If you intend to keep these birds in a large enclosure without restriction, I recommend you take one male and two to three females of a single species. Additional males will sooner or later take off for other locations.

Buy healthy birds with straight toes and a tight covering of feathers. Acquire birds in fall, preferably, especially if you're getting relatively young stock. Don't be tempted to buy just before the start of the breeding season.

Avoid inbreeding as much as possible. In other words, don't mate brothers and sisters; and if so, don't do it several generations in a row. Otherwise you'll run into a lot of trouble, such as crooked toes, degeneration in body size and feather structure, partial blindness, and weakness.

After birds of the pheasant family have ben in a roomy aviary for some time, it is quite likely that breeding will succeed. In the area where young will be raised, you'll need to furnish adequate bedding, such as wood shavings, chopped straw, or commercial matting. Don't let the bedding get blown around. It is to avoid this problem that fine sand and peat moss are not recommended. In addition, chicks tend to eat too much sand, especially if you use heavier sand that is less likely to be blown about. Too great an intake of sand kills the chicks. And peat moss, which seems to get swirled up by the slightest breeze, causes respiratory problems in the chicks. So, pick good bedding and distribute it in a layer 1½ to 2½ inches (4 to 6 cm) thick along the floor.

Once chicks are half-grown, it's worth the trouble to construct a slatted floor in their living space for sanitary reasons. It should be about 14 inches (35 cm) off the ground.

After courtship and mating, the hen will go to her nesting place, usually a secluded spot—under a bush with overhanging branches, for example. This nest will already have been constructed ahead of mating.

Some ornamental fowl will accept artificial nests; these should always be placed in the outside run.

After a hen starts brooding, she should be left alone. Remove the cock from the aviary and house him elsewhere. The same goes for other hens still in the aviary.

Put some small-mesh, woven-wire screens around the brooding hen. Screen sections should be about 1/3 inch (1.10 cm) high and about 6½ feet (2 meters) long. The resulting little pen serves to contain the chicks after they hatch to keep them from getting lost or running away. Put some plantings around the pen for more privacy. Don't have roosts inside the pen. You don't want the chicks to use them, lest they catch a cold. After the chicks are several weeks old, you can remove the pen and the plants around it.

Feed a commercial mix of ornamental fowl poultry feed. Use appropriate mixes in the logical order as chicks grow. Begin with starting mix for the first three to five weeks, and then gradually change over to the hatchling feed, which chicks can eat till they are about six months old. Then use the mix for the next stage, again making the change gradually.

In addition, furnish finely chopped green food, such as chickweed, spinach, dandelions and nettles. Also give the birds finely-ground grit and charcoal. Put each type of food in separate dishes. Once you have birds on hatchling food, you can also furnish them a mix of tropical aviary-bird seed.

Note that newly hatched chicks don't eat or drink the first 24 hours of life.

Furnish fresh drinking water that is not too cold and renew it every day. It helps to put some gravel in the water bowl.

For information on brooding with foster parents and with an incubator, you should refer to specialized literature on the subject.

11

Breeding Ornamental Waterfowl

ORNAMENTAL WATERFOWL include swans, geese and ducks. They are best kept in pairs and always mate on the water, so that a pond is essential. After mating, both partners thoroughly wash themselves. If you plan to breed swans, provide them with at least several square yards of pond space.

Mating begins when the male takes the female by the neck feathers and it takes only about 10 seconds. Swans begin to mate at a relatively young age, but they reach true sexual maturity only much later. For example, black-necked swans can mate at one year of age, but are fully sexually mature only at two or three years of age. A pair of breeding swans should be verified heterosexual, at least three years old, and completely molted.

Species of geese are sexually mature in their second or third year of life. Species of ducks are ready to breed at about one year of age.

Be aware that homosexual behavior is common in swans and other aquatic birds. Two males or two females will build complete nests; the "coupled" females will lay unfertilized eggs. This often happens with captive birds.

Furthermore, many aquatic birds can't be sexed easily because male and female look alike; only among the ducks do we find exceptions.

Sexing requires care and expertise. Van der Mark offers the following suggestions:

Lay the bird to be examined on its back. If necessary, have a helper hold it

This Wood Duck (*Aix sponza*), female, is from North America, and is very closely related to the Mandarin Duck (*Aix galericulata*) of Japan and eastern Asia. Both species are frequently kept and raised in captivity. Female Wood and Mandarin Ducks are very difficult to tell apart.   *Photo by Author*

The Black Swan (*Chenopsis atrata*) comes from Australia and Tasmania, and is known for its curly feathers and red bill. The Dutch discovered the bird in 1697, and took it to Batavia (Indonesia), and thence to the Netherlands.   *Photo by Author*

down. Carefully part the feathers near the genital slit till the opening itself can be clearly seen. Then, peel open the slit very slowly and carefully till the mouth of the intestines is visible. Now, if you see a small, meaty, somewhat rounded rosette, the bird is a female.

Inexperienced sexers often misidentify birds as females because they don't keep the genital opening open long enough. The bird struggles during sexing and the tight muscles of the birds don't relax properly. It's necessary to wait calmly for a while. If the bird is relaxed, then light pressure on the bird's back with the little fingers is enough to get the genital opening completely open.

The bird is a youthful male if during the sexing one can see a partially protruding, light pink-to-white oblong member. The penis of a full grown male can be recognized clearly as a white-to-pink little rod; often it is notched. In swans and geese it easily is as long as a matchstick. Pressure with the thumbs will make it stand out.

Practice makes perfect. It is recommended to practice sexing first on inexpensive birds. In sexing swans, the sexer should watch that the long neck of the swan doesn't come to lie between his feet to be trodden on, because the sexer bends himself completely over the bird during the process.

Small types of geese and ducks are harder to sex because their genitals are smaller. Tree duck species are very difficult to sex because their genital organs lie deeply hidden.

Swans and geese like to have a growth of reeds available as a nest-building site. They also nest under ledges and bushes. Ducks like to nest in the well-known duck baskets. They also will use small coops, especially if these are installed on top of a few stones on an island in the pond. Both geese and ducks often cover their nest with down they pick from their own bodies.

The breeding season usually falls in early spring. Always offer aquatic birds more brooding places than they need to avoid fights. Also, be sure that they have been living in the place where they are to breed for at least six months.

During brooding, which is done virtually always by the hen, the male keeps watch in the immediate vicinity, ready to repel any intruders. While swans brood, they keep building and rebuilding their nest.

You can tell that chicks have hatched by the way the hen seems to sit higher on the nest. The chicks go to reconnoiter their surroundings as soon as their down is dry. They are "nest leavers," as I have said earlier. With the exception of swans and geese, the chicks don't return to the nest once they leave.

In the same food trough you have been using for adult birds, furnish a good, commercial grain mix. Also provide cracked corn, millet, old, soaked bread, and greens—finely chopped grass, duckweed, algae, and other aquatic plants.

After the chicks have shed their down, they can already be removed from their parents. Be aware, moreover, that some drakes can get very annoying and jealous if there are young ducks around. In those cases, house the drake elsewhere.

Regarding notes on wildfowl, I refer the reader to the book by Richard E. Bischop and Russ Williams "*The Way of Wildfowl*" (T. G. Ferguson Publishing Company, Chicago, IL 1971).

12

Brooding and Care by Foster Parents

MANY BIRD FANCIERS keep certain birds for brooding and other birds for laying eggs. Naturally, the "layers" are the more expensive birds. It can easily happen that the layers are degraded into become egg-producing machines. This approach constitutes cruelty to animals and should not be condoned in any way.

When to Use Foster Parents

The proper time to use foster parents is when the natural parents abandon eggs or young. If you don't want to lose them, you have to take quick and effective action.

In principle, you don't need foster parents. Any egg can be hatched in an incubator. But many eggs, those of Canaries and Zebra Finches for example, do cause considerable problems, aside from the fact that it's virtually impossible to hand-raise the hatchlings. Furthermore, an incubator is not always practical because the proper brooding temperature and optimum relative humidity aren't completely known for many cage and aviary birds.

Working with Foster Parents

"Nest leavers," like ornamental fowl, are easier to hand-raise than "nest huddlers," like parrots and finches. For the latter, foster parents are required. Bengalese often serve in that role for exotic finches (as as the Australian types). The Bengalese are completely capable of caring for the eggs and young.

If you keep Bengalese as foster parents, have them specialize in this task. Evidence has shown that couples that have also raised their own young don't always function well as foster parents. Please refer for more details to the earlier discussion of Bengalese and Zebra Finches.

Foster parents ought to be provided with hatchling food and not exclusively with soaked seeds, which Bengalese use by preference to raise their own young.

Note that birds raised by foster parents are sexually oriented to the species of the foster parents, not to that of their own kind. For example, Gouldian Finches raised by Bengalese don't act like Gouldian Finches. Once they reach sexual maturity, they are attracted more quickly to Bengalese than to their own species. This, however, doesn't detract one bit from the fact that Bengalese, along with Zebra Finches, are among the best foster parents for exotic finches.

Larger parakeet species are also frequently raised by foster parents— again preferably by couples that have not raised their own brood. The best success is with foster parents whose brooding period is about the same length, give or take a day or two. The eggs, also should be about the same size as those of the foster parents. Naturally, avoid using birds as foster parents with a history of egg picking.

In raising ornamental fowl of the "nest leaver" type, also match natural and foster parents by size of egg, brooding period, and number of eggs per clutch. Don't give the foster parents more eggs to brood than they would naturally handle, since they wouldn't have enough brood spots for the task.

Among gallinaceous birds, Silkies have gained special fame as foster parents. Be sure to remove the leg feathers from these birds when you start using them as foster parents. The foster chicks would be bothered by the long feathers. Silky hens can even handle eggs that require somewhat longer brooding than their own. They will faithfully and persistently brood eggs for up to 28 days. This opens a lot of opportunities. Place the Silky hens in a separate brooding enclosure, which doesn't have to be particularly large. If there is too little exercise room, release the fowl, preferably around midday, in a separate run. Give them water and food there. Equip the run with a wire mesh floor, so the fowl won't soil themselves with sand and feces.

13

Purchasing Birds

PET OWNERS typically start buying a single Budgie, Mynah bird, Canary or familiar parrot species. Only later does the pet owner start thinking about breeding, and he then acquires a second bird of the opposite sex in hopes it will pair with the first bird. He sets up a homemade breeding cage or he builds a simple aviary in the garden. He may expand to several breeding pairs, inspired by a visit to a local or regional bird exhibition, a bird park or zoo.

Shopping Carefully

The casual approach is far from ideal. Before getting into breeding, first study as many good books on keeping, breeding, and caring for birds as possible. Also read several issues of bird magazines to see where you can most easily acquire the birds and supplies in which you are interested. Fortunately, there are a number of such magazines that are well worth reading, and they have a wealth of advertisements.

Feel free to visit a number of bird stores to look and compare. Talk to the proprietor and be sure to take good notes. Find out what the birds you like are presently eating so that you can offer the same food. (If necessary, you could gradually change to new and better food. I will say more about various diets further on.)

Don't buy birds on order. Remember that no species of bird responds well to a long journey. Always look over the birds you are buying. And don't buy birds during severe cold, because the birds would only get sick from the stress of the change.

Where to buy birds? The question of where to purchase is a matter of mutual trust. Most dealers strive to make their birds as comfortable as possible in their temporary quarters. The cages are light and airy, disinfected, and large enough to keep birds for long or short periods.

As a buyer, realize that dealers have to really work at keeping a collection free of external parasites, given that new stock is constantly being brought in. Similarly, there may be other problems. Observe the birds carefully from a distance. The birds should conduct themselves normally.

What to Look For

The three basic points to watch for with birds is that, first, they shouldn't constantly sit with their head buried in their feathers. Second, they shouldn't be fussing listlessly with their food. And, third, their eyes should be clear.

Also, note the condition of the cages, the food and water dishes. They all should make a neat, clean impression.

In addition to shopping at dealers, consider buying from friends and other breeders. Newspapers and magazines are full of "Birds for Sale" advertisements, especially after the breeding season.

When visiting a private seller, note the same points raised as in examining a commercial establishment. Everything should make a neat impression. The birds should look bright and lively.

Examining the Birds

With commercial and private sale alike, the seller must be willing to let you exchange a bird, if for example, it doesn't turn out to be of the sex you want. You should get a clear, written guarantee to cover points like this.

Do take a close look at birds you consider buying. A healthy bird has a tight covering of feathers, in contrast to sick birds, whose feathers are parted and dull. Look for these signs particularly with the feathers on the bird's head. Feather quality is not so important. If the feathers don't look good, they'll improve after the next molt if the birds are well housed and well fed. However, don't buy birds that pick their own feathers or scratch between their feathers with their toes. Also, don't get birds whose feathers around the cloaca look wet and dirty.

Budgies, especially young ones, should *not* have any missing tail and wing feathers, however. This is a symptom of French molt—an illness which Budgie breeders have termed crawlers, creepers or runners. Some birds will sooner or later get the missing feathers back, but others remain bare in those spots. Some can't even fly!

If you plan to exhibit birds, be sure your purchases don't have missing toes or a deformed beak. Beaks and legs should be free of injury. Dirt should not stick to the body.

Also take a careful look at the breast bone. It's very important that it should not stick out sharply to the front. The eyes should be clear and the bird shouldn't be rubbing them against the roost constantly. Watch out for colds, which claim quite a few victims, especially among recently imported birds that have not yet been acclimated.

Watch the way birds sleep. Very young ones sleep with both feet on the roost, with their head tucked under one of the wings. An adult bird that sleeps this way certainly isn't doing well. Healthy adults sleep with their beaks stuck between their feathers and one leg pulled up to protect it against cold and to conserve energy. Small exotic finches even hunch down, so that the one leg on which they are resting is partially protected by the stomach feathers against the night temperature. Even in the tropics the temperature can differ markedly between night and day.

Bringing New Birds Home

Buy small birds in the morning, so that you will have them home by noon. This way they can spend the rest of the day eating, drinking, and getting used to their new surroundings. Larger birds, such as parrots and pheasants, can be brought home later in the day.

If you transport birds by train or automobile, wrap the transport cage completely in heavy package wrapping paper. Leave the front partially uncovered, however, to let air and light in.

Get newly purchased birds home as quickly as possible. Small tropical birds can't do without food and water very long. Don't put water in the water dish—it would soon spill out. Instead, put a piece of soaked bread in a can to help the birds slake their thirst. Do give them a small dish of seed or put extra seed and moist bread on the bottom of the transport cage. Birds instinctively look for feed on the ground, although there are a few exceptions.

You can always keep seed-eating birds happy with a stalk of spray millet. Hang some spray millet up for the birds when you get home, too.

If you have to bring birds home in the winter—which I don't recommend—avoid placing them in a warm room right away. It literally could kill them. The air in their lungs, air sacs, and hollow bones expands upon being heated. You will understand that this brings about great stress on their body, causing a very painful death.

Instead, get the new arrivals used to a higher temperature gradually. This would be a better method to express your care and concern for cold birds.

Special care upon arrival is needed throughout the year. Birds understandably get shy and scared when they are shaken about in transport cages. So give them absolute rest after you get them home. Leave them alone in their transport cages for a few hours to give them the opportunity to recover.

Do give them lukewarm water and food during that time. I emphasize the fact that the water should be body temperature because the temperature in the transport cage can go up considerably, wrapped as it is in packing paper. Cold water, therefore, could easily bring about illnesses like intestinal problems. Such illnesses should be prevented wherever possible.

Don't put the new arrivals with your existing collection as soon as you reach home. They may be incubating some disease, so a quarantine of about two weeks is absolutely necessary. Furthermore, the new birds can thoroughly upset the existing pecking order in the aviary. Put the new arrivals in a wire cage for a few days when you first put them in the aviary. This way, the new birds and the existing population can safely get used to each other.

After the new arrivals are ready to mix with the birds already in your aviary, watch the new birds carefully for a while. Be fully aware of the goings-on in the aviary. Who is the hell-raiser? Who gets it in the neck? Who needs extra food, more nesting material, or another partner? Don't leave anything to chance and act immediately if anything untoward happens.

Remember, not all birds get along together. Males can upset the whole aviary once they are no longer needed during brooding. And not every species is suited to a communal aviary, especially if it is represented by only two breeding pairs. Zebra Finches, for example, always live embattled lives if two couples are housed together in the same aviary. It's different if there are three or more pairs.

Caring for more than one bird is a real hobby. A well-managed hobby requires time, dedication . . . and money. So, don't overinvest. Buy relatively inexpensive birds to start with and gain experience with them. That way, you don't take too much hay on your fork all at once. Then if you really get caught up in the hobby, you can make the additional investment in the more expensive, harder-to-raise birds.

Acclimating Birds

Acclimating birds is a highly important requirement. The term means getting used to a strange climate, or—by extension—getting used to a completely new environment. Both parts of the definition apply to tropical and subtropical birds moved to a temperate climate. It certainly is not just the change in temperature that causes problems. It also is the stress birds encounter after being caught—an event which, in itself, is quite traumatic.

Before the birds reach you, they undergo all sorts of deprivation in the process—at least in most cases. From the catchers, birds go to buyers,

Sulphur-breasted Toucan (*Ramphastos sulphurata*); approx. 17½ inches (44 cm). Habitat: Mexico (especially the southern sections) to Venezuela.   *Photo by Author*

wholesale exporters and importers, and finally to the retail store and the eventual buyer. Just think about the process! How many times will the birds have been moved to a new cage? How often will they have been exposed to high and low temperatures? How often will they have been fed strange food, and exposed to other stressful situations?

The constant changes are stressful despite the fact that today birds reach their destination far more quickly than in the not-so-distant past. The major difference, of course, is faster transportation, principally by air. Fortunately, international airports cooperate fully in providing as responsibly and as adequately for captive birds as the circumstances permit.

Good importers can make a lot of difference. They can start the birds on food that would be acceptable to the retailer and the bird fancier. Food is not a crucial factor with seed-eating birds and most psittacines, but insect- and fruit-eating birds (the so-called softbilled birds) have a real challenge in adapting to a new diet, particularly if they aren't getting the same insects or fruits to which they had been accustomed in Nature. Therefore, the key to whether such birds have been properly acclimated is whether they eat regularly.

What a bird eats is not all that important—there are many alternatives. The important factor is that the food should be properly digestible. Honey has become a standard transition food. There are few birds, if any, that will refuse honey. It is a good food and it stimulates digestion.

Stimulating digestion is important, because digestion can suffer as a result of transportation. Birds need constant access to food and drinking water because they have a high body temperature and rapid respiration. (During transportation, water can be supplied in thoroughly soaked bread.)

You can see almost immediate untoward results in a bird if intake of food is disturbed in transit for any reason. Behavior changes. It is nervous, moves anxiously through its cage, and pecks at other birds constantly. As the condition worsens, it raises up its feathers, sticks its head under its wings, and goes to sleep in a corner.

Importers should, therefore, constantly work at keeping their birds interested in food. They should be careful to first provide a diet similar to the one the bird ate before capture. The familiar food can be changed to more easily available food, by mixing the two in gradually changing proportions. Variety should be maintained in the menu, meaning a rich assortment of seeds, fruit, and insects.

Importers and other bird dealers also must give a lot of attention to proper temperature. That is difficult, but the good ones do it anyhow—if only to keep their losses at a minimum. Indeed, they have reduced their losses considerably although there is still room for improvement.

The retailer and the fancier himself also will have to heed temperature requirements. Birds should be housed at a temperature between 72 and 80 degrees F (22 and 27 degrees C), in as large a space as possible. This prevents a great loss of body heat in the birds, which is desirable if not essential for proper acclimation.

Birds that won't eat voluntarily despite your best efforts must be force-fed by hand. That is much easier to do with large birds than with small exotics. By the way, be sure to wear gloves if you hand feed large birds, especially psittacines.

In the cage of newly-arrived insect- and fruit-eating birds, cover the bottom with heavy wrapping paper, with the dull side up. Or use newspaper. The cage bottom should be covered with a thick layer of coarse sand for seed-eaters.

At the start, offer food and water in open dishes because the birds have not been introduced to self-feeders or automatic waterers. Drinking water should be room temperature. Don't furnish bath water the first week after arrival. After a week, offer bathwater at room temperature in flat dishes.

Place cages in a sunny, bright location that is absolutely protected from drafts. And remember to keep the birds' surroundings very peaceful and quiet.

Watch new arrivals closely for the first three to four weeks. Then, if all goes well, you can consider moving their cages to an outside aviary. Choose a beautiful, warm day at a time when you can be fairly sure that good weather will hold up for a while. Not surprisingly, the new birds will get a lot of attention during the first few days from the established aviary residents. That doesn't matter though if you leave them in their cages at first. That way the fellow inhabitants of the aviary can get used to them without being able to harm them.

After three days, you can open the cages and let the new arrivals join the others. Spread some food on the floor for the new birds, until they have learned to use the seed dishes. At first, the new birds will still look for food on the ground.

You may want to release new birds in the sleeping shelter. They might prefer to spend the first days there, especially if you spread their food on the shelter floor. This recommendation is less essential if you release birds on a warm day.

It is important to release birds early in the morning, so that they have all day to reconnoiter their new surroundings and find enough possibilities for escaping, hiding, and sleeping. Plantings should be available for the purpose. Many species need special plantings in order to begin breeding in their new quarters.

It is best to put birds in the aviary in the spring, considering the milder temperatures and longer day length, meaning more light for the birds. When autumn approaches, move all (or nearly all) tropical and subtropical birds to a frost-free, if not heated room, preferably a light room with spacious cages in which to spend the winter. Perhaps you can use breeding cages for this purpose. Be sure to separate the sexes. Especially, separate male Zebra Finches, Canaries, and Budgies. Provide artificial lighting, so that it will be light at least 13 to 14 hours per day—for example from 8 AM to 9 PM.

Provide cod liver oil for seed-eating birds, preferably by mixing five drops into about two pounds of seed. Also, regularly provide old white bread soaked in milk. (See the chapter, FEEDING CAGE AND AVIARY BIRDS.)

An outside aviary is best for birds born and raised in captivity, or else thoroughly acclimated. Notwithstanding, *all* tropical and subtropical birds should be placed in lightly heated, large enclosures, at about 60 degrees F (15 degrees C). Outside aviaries can be considered only if the birds have access to an absolutely frost-free night shelter, where they can escape from drafts, rain, and cool night temperatures. During the day, birds could be released into an outside run, provided the weather is good. Late in the afternoon, they should be shooed back into the night shelter.

For your reference, the following species and their close relatives could possibly be housed in an outside aviary if the above conditions are observed:

African Silverbill *(Euodice malabarica cantans)*
Amazon parrots (all *Amazona* species)
Blossom-headed Parakeet (*Psittacula cyanocephala rosa*)
Bengalese (*Lonchura striata,* dom.)
Budgerigar (*Melopsittacus undulatus*)
Chattering Lory (*Lorius garrulus*)
Chestnut-faced Starling (*Sturnis malabaricus*)
Cockatiel (*Nymphicus hollandicus*)
Cuban Finch (*Tiaris canora*)
Cutthroat Finch (*Amandina fasciata*)
Golden-mantled rosella (*Platycercus eximius*)
Gray Singing Finch (*Serinus leucopygius*)
Greater Sulphur-crested Cockatoo (*Cacatua galerita*)
Green Cardinal (*Gubernatrix cristata*)
Green Singing Finch (*Serinus mozambicus*)
Indian Silverbill (*Euodice malabarica*)
Ivory-billed Toucan (*Pteroglossus flavivostris*)
Lesser Sulphur-crested Cockatoo (*Cacatua sulphurea*)
Moluccan Cockatoo (*Cacatua moluccensis*)
Nuns (all species) (*Lonchura* genus)
Olive Finch (*Tiaris olivacea*)
Paradise Whydah (*Steganura paradisea*)
Pekin Robin (*Leiothrix lutea*)

Spice bird or Nutmeg Finch *(Lonchura punctulata)*; approx. 4½ inches (11.5 cm). Habitat: India, Sri Lanka and Australia. The sexes are similar, but the bill of the male is thicker and heavier than that of the female. *Photo by Author*

Silver-eared Mesia (*Leiothrix argentauris*); approx. 7 inches (17 cm). Habitat: eastern Nepal through Indo-China to Sumatra. The female is less vivid in color; her back is grayer, her undersides grayish-white, the red on the throat and rump is missing altogether. *Photo by Author*

Pennant's Parakeet (*Platycercus elegans*)
Purple Glossy Starling (*Lamprotornis purpureus*)
Red Avadat (*Amandava amandava*)
Red-billed Weaver (*Quelea quelea*)
Red-crested Cardinal (*Paroaria cucullata*)
Red-headed Finch (*Amandina erythrocephala*)
Red-rumped Parakeet (*Psephotus haematonotus*)

Java Sparrow.   *Photo by Author*

14

Housing for Birds

IT WOULD BE WRONG to assume that any cage or aviary, indoors or out, is suited for housing birds. Actually, the requirements vary and should be studied before birds are bought. Unfortunately, however, this seldom happens in practice. Notwithstanding, the success of the bird fancy stands or falls on one essential point: whether proper housing is ready and waiting before birds are acquired. That's the golden rule for the breeder.

Housing should be suited to the birds you want to keep. It is true that different species of different sizes can be housed together. Even birds with varying dietary needs can, in some cases, be placed in the same flight, but it is a situation requiring close attention. Otherwise, illness and discomfort will ravage the birds.

Birds can be housed indoors as well as outdoors. There is much to be said for offering your feathered friends fresh air. On the other hand, remember that nothing is as bad for birds as drafts, fog and smog. In response to these conditions, many bird fanciers have created more satisfactory conditions by building housing in garages or basements or in the attic, making use of all the space under the roof. This has worked out quite well, thanks to equipment, such as special lighting, which came on the market in recent years. Attics and unused bedrooms have worked out quite well for housing and breeding canaries and insect- or fruit-eating tropical birds. Apartment dwellers really have no other choice, unless they can get permission to build a small outside flight cage on a balcony or other outside area.

Outside Aviaries

Most fanciers prefer aviaries located in the garden or backyard, however. The model and dimensions of such a facility is a question of personal preference and available space.

Aviaries should always be situated with the front facing south when this is possible. If a front-facing southern exposure is not available, have the aviary facing west. Build with the best materials, so that you won't have to make extensive renovations in just a few years. If you use wood and wire mesh, remember these materials keep going up in price, so you actually save money by building for the future. Some have built flights of chain link; they cost more but last much longer than wood.

Use the narrowest boards possible because these shrink and warp less. For wire mesh, use the familiar hexagonal small-gauge mesh. For large parakeets, however, square welded mesh, which is available with wire of several thicknesses, such as 1.25, 1.47, and 1.65 mm, is preferable. You can buy these supplies in any home improvement center.

Cover the wire mesh of the roof with a second layer of mesh—this one with wider openings, to discourage cats and native wildlife from sitting on the roof and harrassing the aviary birds. Keep a space of at least four inches (10 cm) between the two layers.

Paint the mesh first with petroleum, then with a non-toxic black paint, so that birds will be more visible. Fasten all wire mesh to the outside of the wood, so that birds won't injure themselves. As a further precaution, nail a thin slat along the boards where you have stapled the wire mesh. This protects the birds from all rough edges and fasteners.

A good aviary has three sections: a night shelter, an open run (or flight cage), and a closed run. The shelter should have at least one window, protected with a wire mesh screen. The screen prevents the birds from flying into the window, which would result in broken wings, legs and other injuries.

Protect the aviary against rats and mice. Anyone who has experienced the damage—and especially the fear—caused in the aviary by a night raid by rats will certainly resolve never to let this happen again. Vermin must be kept out. Use concrete flooring or heavy paving stones to keep them from gnawing their way in from below. If you prefer an earthen floor, extend the wire mesh of the walls at least 16 to 20 inches (40 to 50 cm) into the ground. You also can bury glass splinters, broken bottles, and similar rubble 16 inches (40 cm) deep all along the aviary. In a smaller aviary, dig down 16 inches (40 cm) along the whole floor and bury wire mesh along the bottom. Then you can still have attractive plantings in the earth above the mesh.

Another possibility for a small flight cage is to raise the whole aviary floor perhaps 10 inches (25 cm) off the ground, leaving enough crawl space for a dog. My experience is that a good dog will keep rats and mice away. A

Dachshund or small terrier is ideal for this purpose. *Do not use a cat!*

There are rat and mouse poisons on the market that are safe for use around birds. But if children could get to the poison, don't use it.

Be aware that mice and rats can gnaw their way through concrete—with difficulty when it has hardened and with ease while it still is setting. So don't put your entire faith in concrete flooring. Keep the aviary clean. Clear away spilled seed and unused seed left behind by ground-feeding birds like quail and pheasants. And always keep watch for vermin, so you can take prompt remedial action. But prevention is always better than a cure.

In designing an aviary, length is a more important dimension than width. The longer you can make it, the better. Height also isn't crucial, but the flight ought to be at least six feet high. Don't make it too high, however, because if you have to catch your birds—say, to bring them indoors for winter—you're going to have difficulties. For a beginner, an aviary of the following dimensions: five feet wide by 16 feet long by six feet high (1½ meters wide by 5 meters long by 2 meters high). If your plan is to have a breeding aviary for just one couple, then it would be all right to go with dimensions of four feet by eight feet by six feet (1¼ by 2½ by 2 meters).

Every flight or run needs a night shelter that must be lighted to induce the birds to spend the night there or to find shelter there in bad weather. Most birds like to sleep as high up as possible, the well-known roosting position. Therefore, see to it that the roosts in the run are lower than those in the night shelter. This will get the birds to accept the shelter as their sleeping and security area.

If your aviary is long enough, it's wise to cover part of it with waterproof roofing, so that birds can remain outside and still stay dry in rainy weather. There are many birds, however, that like to let themselves get wet during a mild rain shower. This won't hurt them, especially not in spring or summer.

If, for certain reasons, you can't build a night shelter, cover one end of the run on top and on the sides with corrugated plastic siding. This would still give the birds a secure place to be at night and during wet weather.

For an outside aviary especially, build an entry with doors at each end to prevent birds from escaping. Also build hasps for a good padlock. Unfortunately, this is a necessity. With the high prices that most birds bring today, additional protection against theft is also indicated—perhaps an alarm system would be a good investment. In connection with the recurrent incidence of theft, it is wise to band all your birds and to keep a record on each individual. True, it is a small matter for a thief to remove the bands, but prospective buyers tend to be suspicious (and properly so) if they are offered birds without leg bands.

Design the aviary so you'll never run into problems with filling or cleaning the food and water bowls. Choose a convenient spot away from plantings, and you will be glad you did, especially during the breeding season. The ideal method is to build a separate access door from the outside, but that isn't possible in every location.

The floor of the aviary has to be regularly turned over with a shovel or raked loose. In a corner of the aviary, I regularly sow some seeds. As they sprout, they offer a special treat for most birds. Another section is covered, and regularly refilled with a layer of grit.

An uncovered run can be planted with live trees and shrubs. Also landscape the area around the aviary so that it blends in well with your garden as a whole. Your aviary should be in harmony with its surroundings. Choose plants that attract insects. Many will fly through the mesh right into the waiting beaks of your birds.

Don't put live plants in an aviary that houses hookbills, however. They tend to destroy anything they can gnaw at with their destructive beaks. Instead, install some dead fruit trees, willow, or other trees with a good supply of branches.

Indoor Housing

It is possible to construct an aviary indoors or to set up a "bird room," an unused room in the house. To protect the birds, there must be wire mesh screens on the windows. You may want to go beyond this and get reinforced panes. This last is a good safety measure.

It is also of the utmost importance that indoor housing be properly ventilated. Windows, doors, or both need to be situated so that they can be opened without creating a draft—surely the archenemy of all birds.

The principal advantage of indoor housing is that it offers a wealth of options for the breeder. Professionals are known to prefer inside aviaries and rooms over outdoor facilities. Indoor facilities make it easier to check on your birds unless you build several small outdoor aviaries with a very restricted population. Indoor facilities are also good to have when you want to bring birds inside for the fall and winter. Indoor aviaries and bird rooms also appeal to fanciers who don't have a yard or garden and don't want to go into building a glass flight cage.

Breeding boxes in indoor facilities should preferably be located against the far wall. Wooden breeding boxes should not be made of any material in which the layers can separate to make space for vermin. Alternatively, you can install breeding cages. These are often used by breeders of Budgies and Canaries.

If you intend to breed—and have breeding cages—in a bird room, you need to know which sizes and types of cages you will want to use for different species. A random collection of cages will not do.

To breed Canaries for song or for color, you need cages 17½ by 14 by 16 inches (45 by 36 by 40 cm). For other Canaries and Budgerigars, cages measuring 25 by 20 by 17.5 inches (65 by 50 by 45 cm) will work. For all kinds of tropical birds, including Zebra Finches and Bengalese, I suggest 20 by 14 by 17½ inches (50 by 35 by 45 cm), and for love birds and others of like size, cages measuring 33 by 23 by 20 inches (85 by 60 by 50 cm) are needed.

Consider having roosts custom-made of hardwood in various thicknesses to accommodate the different species you will keep. The toes of roosting birds should never be able to close completely around the rod. Also, consider having hand-made nest boxes—in several designs. Use hardwood and other quality material whether you make them yourself or have them built.

For insect- and fruit-eating birds, buy bowls and dishes made of glass because they'll have to be cleansed more often. Consider getting utensils made of hard plastic for hookbilled birds. Dishes for drinking and bathing and for food should be placed so that you disturb the birds as little as possible when you service them.

Indoor housing requires adequate lighting. Natural daylight is best. Fluorescent lighting can be used if natural lighting is inadequate, as in a garage or basement. (I recommend using "Vita Lite," which is available from or can be ordered through your bird dealer or pet supply outlet.) The major advantage of these lights is that they furnish the complete spectrum of natural daylight, including health-promoting ultra-violet rays.

You also may want to plan ahead for artificial heating, even though there are few birds that absolutely require it for survival. However, it's always better to be safe than sorry and furnish a little extra heat. An even temperature also helps promote successful breeding. Heating also enables you to start breeding earlier in the season, even in February or March. A constant temperature of 65 degrees F (18 degrees C) is best. Be careful with heaters, however, and remember that oil heaters, electrical heaters, and other portable units must be used with great care.

The safest units, I believe, are heating elements with thermostats. Those heaters remove humidity from the air, but in such a manner that the area doesn't dry out too much. Air that is too dry can interfere with the hatching process. You will definitely need a hygrometer. If you keep the recommended temperature of 65 degrees F (18 degrees C), the relative humidity should be 65 percent. Keep several large, flat dishes filled with water. If you keep plants in planters or pots, which I recommend, you also will have to spray them with a fogger. Don't worry about wet leaves, because birds generally like to slink along wet leaves to wash themselves.

To keep birds eating as long as they should, turn the lights on for them around 4 PM, especially on dark days, and keep them on till 9 PM. Then, turn them down gradually. Always keep a night light on, however, because birds should not be kept in total darkness. They may fly up at night if they get frightened, and they can bump into something in the dark, hurting

Pelzeln's Saffron Finch *(Sycalis flaveola pelzelni)*; 5½-6 inches (14-15 cm). Habitat: Brazil, Paraguay and Argentina. The hen is brown-gray above and dirty white beneath, with dusky streaks on the breast. *Photo by Author*

Senegal Combassou (Hypochera or Vidua chalybeata amauropteryx); approx. 4 inches (10.5 cm). Habitat: Senegal to Abyssinia. This species is a whydah which does not grow a long tail. The beak is red, the feet salmon to orange. *Photo by Author*

themselves and causing a disturbance among the other birds. Put the lights on again at 6 AM the next day. Get an automatic timer for the lights. It is inexpensive and saves time and trouble. On really dark days, however, make sure the lights stay on all day.

Cages

The basic rule for cages is, the bigger, the better. You could build one yourself quite easily, except for the front section. I like the so-called "box cage" best among commercial models. It has a wire mesh front and the walls, floor, and top are of wood or similar material. Don't use pressed board because it gives off poisonous gases after a while, which hurts the birds' air passages.

You will be able to hang a brooding box on the sides of most cages. Get boxes with removable tops, so you will be able to have a look at the nest, if necessary.

You will need to have several rods for birds to roost on. Arrange them so the birds won't foul one another with their droppings, generally by setting up a "three jump" arrangement. Install roosts so they can't turn. Place them so birds can't hop on but will have to use their wings to reach them.

Cages with vertical bars are best used for Canaries, tropical birds and similar species. Cages with horizontal bars are ideal for parrots and parakeets, which use the bars to move up and down on their "third leg," their strong beak. This type of exercise is good for their health.

Glass show cages are basically "box cages" with a glass front and enough mesh on the top and sides for sufficient ventilation. The pretty scene for which the glass show cage is known is created by properly arranged plants, mostly exotic plants, along with one or two pairs of exotic birds. The cage must be constructed so that the glass plate which makes up the front sits at an angle, with the lower edge slanted outward. This prevents the glass from getting caked and spotted from droppings and splashed water. The sides need to be fitted with several small doors for access to feeding dishes and other utensils.

People who like to keep a few colorful birds will find cages with glass fronts very convenient housing. It is even quite possible to breed birds in them, provided the species you select are among those that are known to respond well to breeding in captivity.

Maintenance

Considerable maintenance revolves around bowls, dishes, roosts and other utensils and furnishings that require constant cleaning to keep germs from multiplying. So take your time in acquiring these. Look at the various models offered in a bird or pet store. Ask what is used for what. At least, you will need good food and drinking water bowls, as well as a flat, earthenware dish for bathwater.

Maintenance is essential if you want to count on having healthy birds that will breed successfully on a continuing basis. Everything has to be kept clean. Automatic feeders and waterers have to be checked regularly, since birds can't do without food or water for long.

Bowls have to be cleaned once a day. For insect- and fruit-eating (softbilled) birds they should be cleansed oftener. Don't use more bowls and other containers than you must—thereby avoiding a messy look. Limit yourself to bowls for water and food plus several troughs for green food.

You can use several sturdy laundry pins to hang green food and other extras at strategic places in the cage or aviary. Cuttlebone is usually sold with a hook, but you can also put a strong piece of wire through it and hang it against the side of the cage or in the covered part of the aviary run. Watch, though, that you don't hang it where birds can be threatened by their natural enemies while they eat.

Regularly check the condition of all wire mesh. Snip off and replace any rusty areas. Check the woodwork also. Remember, many tropical birds are quite small and a hole in roof or wall often suffices as an exit for the entire aviary population.

Wire should be painted with petroleum and then with black or green paint, as was done on installation. This makes wire mesh last for years and makes the birds inside more visible.

Schedule an annual housecleaning. That's the time to thoroughly check wire and wood and to repaint it all. The floor should be dug up at least 12 inches (30 cm) deep and turned over. The nest boxes should be disinfected and put away for the winter. The more attention you give to keeping your bird area clean, the more you'll enjoy both the birds and the facilities in which you keep them.

Cages, especially breeding cages, also need to be cleaned regularly to keep out external parasites. Lice, for example, can cause unsuccessful hatchings. However, it should be clear that during the breeding season there's little or no opportunity to clean a breeding cage.

You will also need a bird net. Get one with a short handle and made of fine mesh or similar material. Catch birds in the air, not against the wire mesh. Don't try to catch all the birds in your collection at once. The birds will get exhausted and accidents will occur. You also can catch birds in the night shelter, but exercise care. Before you start, take out all boxes, bowls, and roosts. Then guide the birds inside quietly, without yelling or waving. Do the job in the morning, so that the birds have all day to recover. They definitely don't like being caught.

Care of Birds

Beginning breeders will do best not to take too much hay on their fork. Starting out with a few easy-to-handle and easy-to-breed species is the best way to proceed.

The bird fancier should avoid housing too many birds in the available space. Overstocking can cause numerous accidents and illnesses. Birds need to have sufficient space for their natural activities. In the breeding season, a bird must be able to establish its own territory, a more or less marked-off area in which it alone is "boss" and from which it will exclude trespassers.

Most birds must be kept in pairs to maintain control over the heredity of the offspring. If two pairs of the same species are kept together, there will be fights and disturbances. I use this rule: either get one pair or three.

One hears many warnings to give special care to newly-acquired, imported birds, but even locally-bred individuals must be given proper attention.

Many species have been domesticated, particularly Canaries, Budgies and Java Sparrows. The Bengalese, a bird bred in the Orient, is really an artificial product and was developed by crossing two species of mannikins, birds which exist in the wild in Indonesia. (The literature isn't clear about the entire origin of these charming birds, which are bred widely and in many color varieties.) People also have developed many mutants of the earlier-named domesticated birds—suddenly appearing variants in color and sometimes in form, like the crested Zebra Finches. One also can consider many representatives of the dove family as domesticated; the Diamond Dove is a good example.

I mention this to show that the bird fancy is a dynamic and progressive activity, augmented by contributions from food manufacturers who have put a selection of excellent foods on the market. Species that couldn't be kept at all, or only for a short time, 20 years ago can be bred in captivity today. Delicate species can be surrounded by just the right temperature and humidity; many fruit- and insect-eating birds don't pose any problem whatever at the present time.

Exotic birds that were caught from the wild overseas arrive in anything but a flourishing condition. Their trip only takes between several hours to two days, but even so, many die en route because of neglect and the stress of being caught and recaught in nets and being moved from cage to cage. Fortunately border control by health and customs authorities includes a good physical checkup, which is good protection against bringing in sick birds.

Most transfers take place in spring, so that birds don't have to cope with northern winters soon after arrival. The birds can then have an opportunity to become acclimated during the summer months.

173

If they are acquired by a conscientious dealer at the point of destination, he will provide proper food and fresh drinking water and allow the birds to recover in spacious cages. After several days, he will also provide some warm bath water for the birds. He will keep the birds in a warm room at about 65 degrees F (18 degrees C) and disturb them as little as possible. Gradually, he will bring the temperature down to the normal outdoor temperature. Then, on a sunny, preferably non-windy day, the birds are put outside.

Most dealers do a good job with this, although they still may lose a bird or two. The survivors, however, usually are strong and healthy and worth buying.

If you buy a pair of birds at this stage, you still shouldn't put them in the outside aviary right away. You do better to keep them inside a while and give them special care.

They should get the best possible food. Almost all birds like to eat insects, and even the so-called seed-eating birds feed their young on insects almost exclusively during their first days of life. Provide the newly-purchased birds with ant pupae, cut-up mealworms, spiders and similar food. Most insects will be quite strange to them, so I have developed a method to get them used to the new fare. I cut open an orange and put the insects on the cut surface. (Or you can hollow out the orange and put the insects in the hollow.) The orange juice gets absorbed by the insects, making their taste much more acceptable. Also consider boiling the ant pupae.

Birds being acclimated get dry or soaked grass seed, spray millet, and a high quality, trusted brand of tropical seed mix. But don't start newly imported birds on seed right away, especially if they are among the small minority that still come by boat. I have noticed many a time that new arrivals overeat, causing constipation. Also don't furnish birds any sand until they are completely accustomed to our climate and food. And take note of the suggestions elsewhere in this book on how to introduce new birds to the existing collection.

To keep birds healthy and lively, maintain a properly-balanced diet and provide adequate housing. In the autumn, put most species in a light room that is draft-free. In winter, be sure to provide a varied diet.

Keep cages and aviaries clean. It's best to check everything thoroughly and dig up the soil on the bottom of the outdoor flights. Replace drinking and bath water daily. Keep feed from becoming musty or spoiled. Pay special attention to artificial and natural roots and to sleeping and nesting boxes.

Always handle and deal with your birds quietly. Don't lose your patience and especially don't go chasing birds that you aren't able to catch quickly.

If you work in aviaries, put on special work clothes. This will keep your good clothes from getting dirty and helps the birds get used to you. Don't keep on entering the aviary in a new, unfamiliar garment—once in a shirt, then in a hat, then in shorts. Birds stay quieter if you're always dressed the same.

Don't ever let friends and visitors inside the aviary. Keep them outside the enclosure, preferably several yards away. Birds just get too upset otherwise. And in the breeding season, a time when absolute quiet is of prime importance, we do best to enter the aviary as little as possible and to keep visitors at a respectable distance.

A special word for fanciers with very limited space. I've seen caged birds kept in living rooms, entry halls, and hallways. This can be done successfully if the birds are guarded from drafts—surely their number one enemy. You can let a couple of caged birds fly loose in the room for a while every day, if you want. Be sure to cover the bottom of cages with a solid layer of river sand or bird grit, which is available commercially.

Pintailed Whydah (*Vidua macrora*); 10-15 inches (25-38 cm). Habitat: Africa, south of the Sahara from Senegal to Eritrea, and also the islands of Fernando Po and St. Thomas.

Photo by Author

15

Feeding Cage and Aviary Birds

COMMERCIAL BIRD SEED can be good—some brands have a well-deserved, worldwide reputation for quality. Commercial seeds are tried out in experiment stations and are chemically analyzed.

Tropical bird seed mixes are important for birds and can be fed with confidence. Some breeders prefer to offer each type of seed separately, and this feeding system has a number of advantages.

Old seeds lose a lot of nutritive value, as experimental analysis has shown. So have fresh seed on hand and don't ever buy more than a year's supply. It's preferable to buy fresh seed monthly.

I always check whether a new lot of seed is fresh by letting it sprout. To do this put a little on a flat plate and add some lukewarm water. Let it stand like that, renewing the water daily. In four days, the seed should be sprouted. If not, change food dealers.

Some special notes are now in order for those who don't want to use mixes and want to provide the various seeds separately.

Canaries

Canaries, especially those kept in aviaries, need a wide variety of seeds. That's understandable since they are free to fly around all day and breathe fresh air. Outdoor living creates appetite. Canaries confined in small cages can't enjoy all this, but that doesn't matter greatly. Most of these cage birds are kept for their musical ability as singers and not for breeding purposes.

The aviary Canary does get used for breeding, which also influences its dietary requirements. The menu should therefore consist of canary seed (the so-called Morocco seed or white seed), rape seed (also called nutsweet summer seed), Hamburg rape seed, niger seed, groats (from oats), hemp, mawseed, lettuce seed, linseed, and plantain seed. The proportions required of these seeds are 35:30:10:8:8:2:2:2:1:2. This basic menu applies to both the aviary Canary and the Canary in a breeding cage.

Canaries in small cages should preferably get canary seed, rape seed and Hamburg rape seed in the ratio of 5:70:25. In addition, furnish some special treats, which are packaged as a special seed mix by several commercial firms. You can also make it up yourself.

In addition to the previous list, also provide some treat mix consisting of canary seed, hulled oats, white lettuce seed, niger seed, and mawseed in a ratio of 50:25:5:15:2. To reach a full 100 percent, add three parts of a special mixture of hemp, plantain seed, and linseed. Especially the true singing Canaries (Harz and Waterslager breeds) love this special mix. You can actually hear the difference in their singing. If singing Canaries sing too quietly, increase the percentage of the treat mix, but temporarily take away the rearing and conditioning food, strength food, and white bread soaked in milk.

Also provide green food, such as lettuce, cabbage, chickweed, Brussels sprouts, and finely chopped carrots. And give them cuttlebone, universal and concentrate food, ant pupae, and an assortment of fruit.

Tropical Seed-Eaters

Let us start with small exotic birds, such as Zebra Finches, Bengalese and Red Avadats. Limit them to Senegal millet, La Plata panicum millet, canary seed and niger seed in a 80:5:10:5 ratio. This ratio, of course, refers to a mix. I prefer to feed each type of seed separately for all tropical seed-eaters.

For bigger species, consider La Plata panicum millet the staple food. Provide some canary seed and Senegal millet as well, in a ratio of 60:25:15—if you decide to make a mix.

Don't pay credence, however, to the old canard that all you have to provide tropical birds is millet. You have to provide the variations suggested to maintain the birds' health.

Further, be sure to regularly furnish the birds some spray millet and a fresh supply of green food daily. Also provide some insects—if at all possible at least twice a week. Then give them some old bread soaked in milk or water, especially during the breeding season. Other excellent food is mosquito larvae, water fleas, tubifex, mealworms, ant pupae (sometimes

178

erroneously called "ant eggs"), and fruits, especially juicy ones. You also can supply universal food and—during the breeding season—some hatchling food of the familiar commercial brands.

Insect-Eaters

It is difficult to give a definitive menu for insect-eating birds, which require a lot of individual attention. The care of these birds really should be left to thoroughly experienced bird fanciers.

Many species have all but solid droppings, which requires special attention for the floor of aviaries and cages. Ordinarily, this means putting a new floor covering, like heavy wrapping paper covered with beach sand, in their cages every day. In aviaries, this means digging up the ground thoroughly at least twice a week.

A number of food products for these birds is available commercially, but they have to be supplemented with insects. Small or cut-up mealworms, enchytrae (especially during the breeding season), ant pupae, grasshoppers, flies, spiders, beetles, earthworms are good. Avoid any from ground where artificial fertilizer has been spread. Meat or fish grubs that have been washed carefully can also be offered, but there shouldn't be any spoiled bits of meat sticking to them. In addition to insects, the birds should get a wide variety of fruits, especially juicy ones.

Fruit-Eaters

The comments made about cage and aviary maintenance for insect-eaters also applies to the fruit-eaters, only more so.

They can be fed all kinds of fruit: pears, apples, grapes, soaked raisins and currants, cut-up bananas and grapefruit are all accepted. Don't forget to also furnish an orange each day, cut across the sections. Impale the fruit on nails driven through a small board and hang it in the aviary. Birds also should be given universal food of top quality.

Fruit eaters also will gladly take applesauce laced with sugar and universal food. In autumn, consider feeding bird-cherries and elderberries. Branches of these can be put into the aviary and renewed periodically.

If you feed fruit with stones or pits, remove these first.

Recently imported birds should be given an orange cut into quarters and covered with universal food. The new arrivals can become acquainted with unfamiliar food in this way. They get accustomed to it rather quickly, which is convenient if the bird keeper occasionally runs low on fresh fruit.

Always cut larger fruits into two or more parts. Put these on separate feeding boards, so that all birds will have access to a full selection of fruit. This prevents unnecessary fights for favorite foods.

Parakeets

These hookbills are not nearly as finicky in their diet as is generally supposed. Still, adequate attention to their menu is imperative to maintain truly healthy birds.

Smaller species, like Budgies and love birds, can be given a mix of canary seed, white millet, La Plata panicum millet, standard millet, hemp, and hulled (but not broken) oats, in a ratio of 25:20:30:10:5:10.

The same mix can be given to Cockatiels and larger parakeets, like Ringnecks, Blue-wings, and Golden-mantled Rosellas. Also add a mix of sunflower seeds and peeled oats. For these birds, also, furnish the required seeds separately rather than in a mix. Extra grass seeds and oats are recommended, especially for the larger species.

In addition, furnish rearing and conditioning food, concentrates, unroasted peanuts, and fruit, like apples, pears, raisins, cherries and bananas. Also provide insects, like ant pupae, enchytrae, and mealworms.

Larger species of parakeets get eight to 10 mealworms per day, but not more! Smaller birds get one or two. In the breeding period, cut up the worms (so that they can be fed to the young) and "cook" them by holding them in boiling water for several minutes. Old nylon hose are handy for the purpose.

Further, give green food, like lettuce and sprouted seeds. And don't forget to furnish grit and cuttlebone.

Larger parakeets and most parrots must be fed from an open bin, which is a problem, because these birds spill a lot of feed. To minimize waste, don't fill feed bins to the rim, but rather, only about half full. The bin itself should be placed in a large, flat, earthenware dish, so that spilled seeds can be recovered. Clean out dirt and hulls from the spilled seed, and you can put it back into the feed bin.

Put rearing food in porcelain dishes, or those made of glazed earthenware (like small food dishes for baby chicks). Supply a relatively small amount of rearing feed, because it tends to spoil quickly. Place the containers in a covered area of an outside aviary for the same reason.

Parrots

Be sure that parrots eat a varied diet. If left alone, parrots will eat only sunflower seeds and leave their other food untouched. You need to act resolutely and force the parrots to eat the other food. Besides the sunflower seeds, they should get La Plata panicum millet, canary seed, unpeeled or unhulled oats and hemp in a ratio of 45:15:35:5. Also give them some peanuts in the shell.

You may wonder why I am so opposed to feeding only sunflower seeds. I have to admit up front that parrots can live on these seeds in good condition for years. But sooner or later, deficiencies will arise, resulting in a slow but certain loss of health. You may notice it in a heavy feather loss, especially on the wings, tail, breast and lower body. By the time these symptoms are noticed, it will take considerable effort to bring the birds back to a normal condition. Worse yet, the birds are now accustomed to eating only sunflower seeds. They often don't want to touch other types of seeds, so that they eventually die from a vitamin deficiency.

Never feed parrots any cake, candy or other sweets. Zoos and bird parks have explicit signs warning visitors not to feed sweets to birds, and for good reason. Such "goodies" can make them sick.

SPECIAL DIET COMPONENTS

Green Food

The easiest and best way to provide green food to aviary birds is to sow some seeds in a sunny corner. It can be the same seed as you put in their regular diet. The birds just love the sprouted seed.

If possible, sow the seeds under the regular food bin. This way, spilled seed will fall in your "green food" patch and so help maintain it the whole season.

Never feed vegetables that have been sprayed with chemicals; these could kill all your birds. Always wash green food under running water, just as a precaution.

Other items you can furnish are chopped boiled eggs and grated carrots. They are excellent as extras for the birds, especially for those with red feathers. But don't overfeed these extras; too much can cause liver problems. Other good sources of carotene are young spinach, thistle seed, and young wild nettles—particularly the tops of these weeds. Birds also love a little chickweed *(Stellaria media)*, which grows wild in many places.

Egg Food or Concentrates

This food is made commercially. It needs to be constituted with care, so only buy the best brands that are sold worldwide. And don't overfeed. Concentrates are rich in carbohydrates, fats, and several vitamin complexes.

You can make up concentrates yourself, if you take special care. Fry a beaten egg in a little butter, and don't let it burn. Cut the omelet you have made into small pieces or mash it with a fork. Let it cool.

Meanwhile, crumble up four and a quarter slices of rusk with a rolling pin. Also, rasp a carrot, about four inches long. Then add a finely grated apple, or applesauce. Mix these ingredients into the omelet, preferably with the juice of half an orange. Don't make up more concentrates than the birds can use up in a single day.

It's best to offer concentrates in a porcelain dish, or one of glazed earthenware, or galvanized steel. To avoid spoilage, put out small quantities of concentrates at a time and protect the food from the weather. Place it in the covered part of the aviary or put a little roof over the feeding station.

Spoilage really is a great threat in any food. If you notice the slightest sign of spoilage, throw out the old food and offer new. Otherwise you'll have sick birds; be especially careful with the food you give to hatchlings.

Bread

I recommend giving birds a slice of stale bread soaked in water or milk, especially during the breeding season. You can use white bread or brown, but white is better. With this food, also, guard against spoilage. Don't give the birds more than they will eat in a day—and not *every* day. Three times a week is the maximum. Outside the breeding season, reduce the bread ration to only once or twice a week.

Cod Liver Oil

Birds need certain vitamins, but don't provide too much. Add about five drops per two pounds of seed and mix well. Let the mix stand a day before feeding it to birds.

Be careful not to oversupply cod liver oil. Keep a balance. It is needed, especially during the breeding season, but giving too much is harmful. The same is true for certain vitamins and vitamin combinations. Don't oversupply them. If you feed the right diet, you may not have to add vitamins at all. If you do use them, be sure to store them in a dark, cool place to preserve their potency.

Water

Water is needed for drinking and bathing. Furnish rain water for both purposes if your water supply is high in calcium or other minerals. If air pollution in your area makes using rain water unwise, bottled spring water should be provided to your birds.

Parakeets and parrots drink and bathe little. Instead of bathing, they prefer to roll around in wet grass or let themselves be wet by the rain. They also like to get a spray shower from the garden hose. Do this faithfully every week—weather permitting. Use very low pressure!

All other birds like to bathe, much like people. Provide enough bathing dishes, so all birds have good access. Drinking and bath water must be changed at least once daily, and more often on warm days.

Remove bathing dishes as soon as you expect frost. Don't allow wet feathers to freeze, because it means almost certain death. If your drinking

Green-backed Twinspot *(Mandingoa n. nitidula)*; approx. 4½ inches (11.5 cm). Habitat: Mozambique. The female resembles the male, but has a pinkish colored face; the breast is buff gray.   *Photo by Author*

Scaly-crowned Weaver (Sporopipes squamifrons); 4 inches (10 cm.). Habitat: southwest Africa. The sexes are alike.   *Photo by Author*

dishes are big enough for bathing, install some wire or mesh over them to keep birds out.

The best waterer is a commercially-made, rather expensive rock with a trough for the water to flow through. I like it because birds can't get into it, nor can dust and dirt. Regular fountains and automatic waterers also can be fouled, but generally less than open dishes.

Honey water is an excellent tonic for exotic birds. I prepare it by dissolving a tablespoonful and adding this to about 1½ quarts of drinking water. During the breeding season, I like to also add a tablespoonful of evaporated milk to the drinking water.

For bathing, use one or more shallow earthenware dishes. Place them on flagstone or tile, away from roosting or sleeping areas, so that the water will be contaminated as little as possible by droppings, sand and dirt.

Sand Bath

Birds also like to take sand baths. They are effective against lice and for rubbing off new feather sheaths at molting time. Install fine sand in a dry, clean spot in the aviary, or put dry, fine sand in shallow earthenware dishes. The birds will disover and make good use of it.

Cuttlebone and Grit

The shell or exoskeleton of the cuttlefish, technically known as cuttlebone, which it sheds periodically, is the recommended source of calcium for captive birds. It is quite important for all birds, especially young ones. Be sure cuttlebone is available in the breeding season and at molting. It is simple to feed and is available, with a hanger, in every good bird store. Hang it in the covered part of the aviary or under a special little roof so it stays dry.

You can use cuttlebone that you gather on the beach, but if you do, soak it for several days to bring down the salt content.

Grit also is of importance. It contains calcium, too, as well as charcoal, iron, magnesium, and iodine. (Cuttlebone, with the same constituents, also contains important salts.) Grit is important for proper digestion and bone growth. It also is available commercially.

Roosts or Perches

Roosts must be of two kinds: ordinary rods for sitting and swings or sticks for amusement.

Sitting rods should never be too thin and could be round—although they may be flattened out on top. They can be swinging or immovable and made of hard wood. The swinging roosts are "playthings"—fun, but not essential. Immovable roosts are absolutely essential for birds that are courting.

184

Thin roosts are purely for diversion and generally consist of natural vegetation, which, if formed and placed properly, also provide natural nesting places. Of course, birds will use these roosts for sitting or sleeping, too, particularly in the summer. Still, I recommend you provide separate sleeping roosts.

Install these roosts in the inside aviary or the covered portion of the outside aviary. Use sturdy, thick rods of hard wood. I emphasize the wood must be hard and free of bark to keep lice out. I also emphasize *thick* rods, because the toes of the birds should not completely encircle the roost if the birds are to get proper and thorough rest. If the thickness is right, furthermore, the birds' nails won't grow too long, especially if we take care to vary the thickness of the different roosts we provide. And the muscles of the birds' feet will remain supple.

Put roosts up high. Birds like to sleep as high up as possible. Be sure there is enough roosting space to avoid fighting at "bedtime." And don't install roosts above each other, because otherwise birds on top foul the birds below with their droppings. In outside aviaries, also furnish some wind- and draft-free sleeping places, such as several half-open small sleeping shelters.

Nesting Places

The most important aspect of nesting places is that they should be present in abundance. Each pair of breeding birds should have a variety of options. Only if they have a wide selection of potential nesting places will birds start nesting and breeding.

The types and sizes of nest boxes required by aviary birds are listed in an earlier chapter, NATURAL BIRD BREEDING. Still, some additional remarks are in order.

First, don't assume that birds will always use the nest boxes recommended for their species. I have observed a supposed "tree hole brooder" that happily made an open nest or used a half-open nest box. And "semi-hole brooders," like Zebra Finches, will sometimes select an open nest, like a coconut half, or a completely closed-in place, like an entire coconut, despite a variety of available sites that it supposedly prefers. Don't be surprised if they change from one to the other in the same breeding season. I have seen this happen a number of times.

It is precisely because birds do surprise us with their preferences that we need to anticipate the choices and make the options as broad as possible. Count on providing at least twice as many nesting places as you have breeding pairs.

There are natural and artificial sites.

Natural sites are offered by vegetation. In selecting shrubs for the aviary, pick types that form dense branches with forks at several places. If you have shrubs that have too few good branches, develop these by cutting the shrubs back to a few promising buds. Meanwhile, you can cut some branches from another tree or bush and tie them together with wire to form a good nesting place. For ground-nesting birds, provide reeds, rushes, and low bushes, such as heather. Such vegetation is an integral part of a good, outdoor aviary of the right size.

Artificial nesting places fall into three basic types:

1. Nest boxes—open, half open, and enclosed models. Nesting logs of birch or beech are really "closed" nest boxes.

2. Nest baskets—coconut shells, woven baskets of various types, bins, and bowls.

3. Miscellaneous—cork, heather, woven straw and similar materials. These are for specialized uses.

Don't begin offering artificial nesting places until early spring and remove them after the breeding season ends. Keep them clean and sanitary. In the winter, give them a thorough cleaning. Then sanitize them inside and out with carbolic acid or stain. Remember, your goal is to produce totally healthy young, and nesting places and nests that are infested with vermin or infected with germs would interfere with that goal.

GUIDE TO WOODEN NEST BOXES

Many birds will accept wooden nest boxes as appropriate places to lay and brood eggs and to raise their young. The following guide provides dimensions for nests for a wide variety of bird species. If the species you are interested in is not named herein, use the dimensions of a similar-size bird to guide you. The wood used in the nest box should be 3/4" to 1" thick. The measurements given are inside measurements, the first being inches and the figure in parentheses ( ) being centimeters.

Species	Length	Width	Height	Entrance Hole
Bengalese, Zebra Finch, other domesticated small finches	5 (12)	5 (12)	5 (12)	1¼ (3)
Budgerigars	6 (15)	6 (15)	8.(20)	1½ (4)
Love Birds except Peach-faced	6 (15)	6 (15)	11 (28)	2 (5)
Peach-faced Love Bird	7 (18)	7 (18)	11 (28)	2½ (6)
Golden-crowned Conure, Tovi Parakeet	8 (20)	8 (20)	12 (30)	2-3 (7)

186

Species	Length	Width	Height	Entrance Hole
Cockatiels, Bourke's Elegant Grass, Turquoisine Grass Parakeets	10 (25)	10 (25)	12 (30)	2-3 (7)
Catharina Parakeet, Cactus, Jendaya, Nanday, Painted, Red-bellied and White-eared Conures	10 (25)	10 (25)	14 (35)	3¼ (8)
Blue-fronted, Festive and Yellow-fronted Amazons	14 (35)	20 (50)	16 (40)	5 (12)
African Grey Parrot	10 (25)	10 (25)	20 (50)	4¼ (11)
Ringneck Parakeets	10 (25)	10 (25)	14 (35)	3¼ (8)
Plum-headed Parakeet	8 (20)	8 (20)	12 (30)	2¼ (6)
Rosella species	10 (25)	10 (25)	14 (35)	3¼ (8)
Crimson-winged Parakeet	12 (30)	12 (30)	16 (40)	4¼ (11)
King Parakeet	10 (25)	10 (25)	14 (35)	3¼ (8)
Alexandrine Parakeet, Green-naped Lorikeet	12 (30)	12 (30)	18 (45)	4 (10)
Barnard's Parakeet, Port Lincoln Parakeet	10 (25)	10 (25)	16 (40)	4¼ (11)

Final Thoughts on Maintenance

Effective maintenance presupposes good facilities. A well-designed aviary has as few corners and sides as possible. It is finished off with carbolic acid and stain inside and out. The floor is of fresh, natural sand or is covered with clean river or beach sand.

Good maintenance, first, requires that the sand on the floor be freshened repeatedly and that the floor itself is turned over with a spade. The frequency of this chore depends entirely on the size of the aviary and the density of the population. Guide yourself by the principle: Never have a dirty floor; better too much work done than too little.

Second, every spring (at the very least) clean and sanitize all roosts, sleeping shelters, hiding places and other furnishings you supply. Natural vegetation needs to be cut back to remove rotting and dying wood. Then clean off everything with spray from a garden hose. Replace plants as needed.

Third, maintain the aviary as a whole. In spring, move the entire bird collection into cages long enough to thoroughly clean, paint, and service the whole aviary inside and out. Check the roof for leaks. Find and seal cracks that promote drafts. Repair tears in the wire mesh. Maintain and make any repairs on locks and hinges. In short, check and restore everything. Done every year, preventive maintenance avoids nasty surprises during the coming season and preserves your collection as well as your property.

Fourth, keep up daily chores. When you replace and refresh feed, drinking water, and bath water, clean all the dishes, bowls, and fountains. Automatic waterers and self-feeders should be cleaned once per week.

Altogether, maintenance is a year 'round chore, with a major spring cleaning—and a second major overhaul in the fall, if you have birds wintering in outdoor facilities. Inside facilities, of course, should be equally well maintained—including cages, glass show cages and all your bird-related equipment.

Star Finches *(Poephila ruficauda)* are from Australia and are a pleasing addition to any collection of birds. In this photo the young bird is perched above his parents.   *Sloots*

Appendix

Major Publications

Most Avicultural Societies publish magazines, bulletins or newsletters for their members. In addition the following national publications are of interest to all bird fanciers:

American Cage-Bird Magazine (monthly)
3449 North Western Avenue
Chicago, Illinois 60618 (USA)
(includes a monthly directory of bird societies)

The A.F.A. Watchbird
American Federation of Aviculture Inc.
P.O. Box 1125
Garden Grove, CA 92642 (USA)

Avicultural Bulletin (monthly)
Avicultural Society of America, Inc.
734 North Highland Avenue
Hollywood, CA 90038 (USA)

Bird Talk (bi-monthly)
P.O. Box 3940
San Clemente, CA 92672

Bird World (bi-monthly)
11552 Hartsook Street
No. Hollywood, CA 91601
Mailing address:
P.O. Box 70
No. Hollywood, CA 91601 (USA)

Cage and Aviary Birds (weekly)
Surrey House
1, Throwley Way
Sutton, Surrey, SM1 4QQ (England)
(Young birdkeepers under sixteen may like to join the Junior Bird League. Full details can be obtained from the J.B.L., c/o Cage and Aviary Birds.)

Major Societies

Australia:

Avicultural Society of Australia
P.O. Box 48
Bentleigh East
Victoria

Great Britain:

The Avicultural Society
c/o Mr. H. J. Horsewell
20 Bourbon Street
London, W1

Canada:

Canadian Avicultural Society, Inc.
c/o Mr. E. Jones
32 Dromore Crescent
Willowdale 450
Ontario, M2R 2H5

Canadian Institute of Bird Breeders
c/o Mr. C. Snazel
4422 Chauvin Str.
Pierrefonds, Quebec

New Zealand:

The New Zealand Federation of Cage Bird Societies
c/o Mr. M. D. Neale
31 Harker Street
Christchurch 2

United States of America:

Avicultural Society of America
c/o Mrs. Doris Mayfield
6606 Enfield Avenue
Reseda, CA 91335
(International association)

American Federation of Aviculture
P.O. Box 327
El Cajon, CA 92022
(Dedicated to the conservation of bird wildlife through public education, encouragement of captive breeding, scientific research, and monitoring legislation affecting aviculture—National association.

National Cage Bird Show Club, Inc.
c/o Mrs. Margie McGee
25 W. Janss Road
Thousand Oaks, CA 91360
(National association)

Bibliography

Bates, H. and Busenbark, R., *Finches and Softbilled Birds,* TFH Publications, Neptune, New Jersey.

Idem: *Guide to Mynahs,* TFH Publications.

Idem: *Parrots and Related Birds,* TFH Publications.

Binks, G. S., *Best in Show—Breeding and Exhibiting Budgerigars,* Ebury Press and Pelham Books, London, England, 1977.

Clear, V., *Common Cagebirds in America,* Bobbs-Merrill, Indianapolis, Indiana, 1966.

de Grahl, W., *Parrots,* Ward Lock, London, England, 1981.

Dodwell, G. T., *Canaries,* TFH Publications.

Freud, A., *All About The Parrots,* Howell Book House., New York, New York, 1980.

Gallerstein, G. A., D.V.M., *Bird Owner's Home Health and Care Handbook,* Howell Book House Inc., 1984.

Hayward, J., *Lovebirds and Their Colour Mutations,* Blandford Press, London, England, 1979.

Immelman, K., *Australian Finches,* Angus & Robertson, Sydney, Australia and London, England, 1972.

Low, R., *Parrots, Their Care and Breeding,* Paul Elek, London, 1977.

Petrak, M. L. et al., *Diseases of Cage and Aviary Birds,* Balliere-Tindall, London, England and Lea & Febiger, Philadelphia, Pennsylvania, 1969.

Restall, R. L., *Finches and Other Seed-eating Birds,* Faber & Faber, London, England, 1975.

Rogers, C. H., *Encyclopedia of Cage and Aviary Birds,* Pelham Books, London, England, 1975.

Rutgers, A. (ed. K. A. Norris), *Encyclopedia of Aviculture,* Blandford Press and British Book Centre, New York, New York, 1970, 1972.

Scoble, J., *The Complete Book of Budgerigars,* Lansdowne Press, Sydney, Australia, 1981.

Vince, C., *Keeping Softbilled Birds,* Stanley Paul, London, England, 1980.

Vriends, M. M., *Encyclopedia of Lovebirds,* TFH Publications, 1978.

Idem: *Encyclopedia of Softbilled Birds,* TFH Publications, 1980.

Idem: *Handbook of Canaries,* TFH Publications, 1980.

Idem: *Handbook of Zebra Finches,* TFH Publications, 1980.

Idem: *Popular Parrots,* Howell Book House Inc., 1983.

Idem: *Starting an Aviary,* TFH Publications, 1981.

Idem: *The Complete Cockatiel,* Howell Book House Inc., 1983.